BIM: Contractual and Legal Perspectives

BIM: Contractual and Legal Perspectives

Peter Barnes

Published by Emerald Publishing Limited, Floor 5,
Northspring, 21–23 Wellington Street, Leeds LS1 4DL.

ICE Publishing is an imprint of Emerald Publishing Limited

Other ICE Publishing titles:
Automation and Manufacturing for the Built Environment.
Steve Thompson. ISBN 978-1-8360-8599-7
Essentials of Digital Construction: Lessons learned from digital transformation.
Amador Caballero. ISBN 978-1-8354-9446-2
BIM in Principle and in Practice.
Peter Barnes. ISBN 978-0-7277-6369-3

A catalogue record for this book is available from the British Library

ISBN 978-1-83608-515-7

Cover photo: John_T/AlexanderTrou/Connect Vector/Shutterstock.com

Commissioning Editor: Michael Fenton
Content Development Editor: Cathy Sellars
Production Editor: Benn Linfield

Typeset by KnowledgeWorks Global Ltd
Index created by David Gaskell

Contents

Preface

Building information modelling (BIM) is having an increasing impact on many areas of operation in the construction industry, both in the UK and further afield.

People are becoming more familiar with the BIM process and better understand that BIM is not something to be feared but is a technological development that should be welcomed by all in the construction and civil engineering industries. Of course, there is a learning curve, as there is with all new technologies, but rewards for becoming immersed in the BIM process are well worth the effort.

While the technological advances are leaping ahead, some of the more difficult legal, contractual and liability issues are lagging behind somewhat.

Presently, the majority of participants in the construction and civil engineering industries that are operating BIM use a standard form contract, with a BIM protocol document annexed to it. However, it is not always clear what the legal standing of the BIM protocol document is, nor where it comes in the hierarchy of contract documents. There is also often uncertainty about where the risks, including the design risk, lie and it is not always clear which party carries the liability of failure and/or is required to hold the insurance cover for that liability.

At this still relatively early stage of the development of BIM, this is not a major concern, but, as BIM takes more and more of a hold on the construction and civil engineering industries, it will become one of the major issues to be resolved and, if it is not resolved, this will almost certainly hamper the progress of and the adoption of BIM.

It must be remembered that, apart from the technological advances that BIM brings with it, a parallel and important side benefit of the use of BIM is the fact that the participants in the process (clients, consultants, designers, contractors and sub-contractors) will all need to collaborate with each other. Therefore, after all the many years of talking about how a collaborative behaviour pattern can be encouraged and applied, it may be a technological advance (i.e. BIM) that will turn out to be the long sought-after catalyst for true collaborative working in the construction and civil engineering industries.

Against the above background, and particularly the continually fast-developing BIM process, this book is intended to assist parties in recognising the contractual and legal obstacles that may occur in respect of the BIM process, and also in guiding them as to how those obstacles may be dealt with, so that the full benefits of BIM may be achieved.

It is hoped that the readers of this book will find it to be informative and easy to read and understand, and that it will develop an interest in them to investigate how the legal and contractual aspects surrounding BIM may be best dealt with.

Acknowledgments

I am most grateful to all those who have contributed to my knowledge about the subject matter of this book. Also, I am pleased to give my special thanks to my wife for the unstinting support and encouragement she has given to me while I have been writing this book.

About the author

Peter Barnes MSc DipICArb FCIArb FCIOB MRICS MICE MCInstCES has been actively involved in the construction industry for over 45 years. He is a Chartered Arbitrator, a Registered Panel Adjudicator, an Accredited Mediator and a Chartered Conservationist. He started as a Quantity Surveyor and progressed through to being the Chief Quantity Surveyor for a London-based building contractor. In addition to this, he also became the General Manager of the construction arm of the same company. Since moving into consultancy work 25 years ago, he has been appointed to act as arbitrator, adjudicator, mediator and expert witness in respect of both construction and infrastructure claims and disputes. He provides seminars and training on a wide range of contract and commercial topics relating to the construction and infrastructure industries, and he gives training on JCT contracts and sub-contracts on behalf of the JCT. He is a member of the Adjudication Society, the King's College Construction Law Association, the Society of Construction Law, and is a Liveryman of the Worshipful Company of Arbitrators. He is the author (and/or co-author) of many published books, including *The JCT 05 Standard Building Sub-Contract; Subcontracting under the JCT 2005 Forms; Delay and Disruption Claims in Construction* (now in its third edition); *BIM in Principle and in Practice* (also now in its third edition); and *BIM for Project Managers*. He has also had many articles published in various construction industry/legal magazines.

Peter Barnes
ISBN 978-1-83608-515-7
https://doi.org/10.1108/978-1-83608-512-620251001
Published under exclusive licence by Emerald Publishing Limited

Chapter 1
Introduction

The acronym BIM stands for building information modelling or building information management.

BIM is a computerised digital form of construction and asset operation which brings together technology, process improvements, data collection and digital information. It applies to all sorts of construction, civil engineering and infrastructure assets, whether that relates to new build, renovation, refurbishment or maintenance projects.

At its most basic level, BIM is a process that is used to design, understand and demonstrate the key physical and functional characteristics of an asset on a 'virtual' computerised model basis, providing the opportunity to concurrently design and visualise the asset in three dimensions (3D).

At its more advanced levels, BIM is the use of a computer software model (i.e. a digital representation) to simulate the construction of and the operation of an asset throughout its entire lifecycle.

The resulting computer model makes it possible for views and data appropriate to suit various users' needs to be extracted and analysed, generating information that can be used to make decisions which can improve both the process of delivering the asset as well as its use during its entire lifecycle, both as an entity in its own right and also (in some cases) as it is incorporated into its surrounding location and environment.

BIM is now widely adopted, both in the UK and internationally, and those that use it are aware of the many benefits that BIM brings with it, including, among other things, reducing the amount of inaccurate and conflicting information, saving time and money, saving natural resources and reducing the carbon footprint of a building.

Further, whereas documents such as drawings, schedules and specifications were previously written and stored in hard copy form, this information is now being mostly digitally generated and stored on computer servers and/or on cloud servers which can be remotely accessed.

As time moves on, and as technology improves and develops, BIM models will give more and more direct access to the stored construction documents through links to the appropriate server and, even now, the digital flow of information from inception through to demolition is becoming more and more common.

Technological improvements are occurring frequently, but the contractual and legal consequences of using BIM are (almost inevitably) lagging behind somewhat; the purpose of this book is to highlight the legal and contractual issues that need to be taken into consideration when operating under BIM, to illustrate the ways in which these issues are presently addressed and to consider how the issues may be further addressed in the future.

Several construction and infrastructure industry reports (including *Digital in Engineering and Construction* (Gerbert *et al*., 2016); *Rethinking Productivity across the Construction Industry* (Lara, 2015); *Modernising Construction* (UK NAO, 2001)) have identified present systemic problems in the construction and infrastructure processes emanating from the poor levels of collaboration; the matter of collaboration (a key component of the BIM process) is considered later within this book.

A major aspect of BIM is the potential full collaboration of the entire project team – the employer, the architect, the engineers, the consultants, the contractor and the specialist contractors – in developing the project design. This full collaboration (which has been sought for many years) not only allows for the increased speed of project delivery, enhanced economics for the project and true lean construction at all levels but also has the potential to change the relationships between the participants in the construction industry, from the more traditional contracts based on obligations and rights to the more modern partnering and collaboration associations based on a fair allocation and sharing of risks and liabilities.

Navigating the legal and contractual landscape in BIM is essential for the successful and ethically sound implementation of projects within the construction industry. Addressing legal issues related to data ownership and liability is a basic step that requires clarity and precision in defining responsibilities among the various parties involved in the process. Clearly, by establishing unambiguous protocols for the ownership of data and outlining liabilities associated with the accuracy of BIM-derived information, potential disputes can be mitigated, thereby fostering an environment of trust and collaboration.

Therefore, the updating of forms of contracts to accurately reflect BIM requirements and responsibilities is vitally important. Further, as BIM methodologies continue to evolve, contractual frameworks must be agile, adaptable and responsive to the changing demands. Clarifying and establishing roles and responsibilities within contracts not only removes ambiguity but also facilitates effective communication and coordination among the many participants.

Ensuring compliance with relevant industry standards and regulations is plainly a non-negotiable aspect of BIM projects. Adhering to established standards not only promotes consistency and reliability across projects but also safeguards against legal complications. By aligning contracts with industry standards, BIM participants contribute to an environment that not only adheres to best practices but also fosters innovation and ethical conduct within the construction and infrastructure industries.

Exciting times lie ahead, and we are still at the very early stages of that new journey, but to make sure that the journey proceeds unhindered we must ensure that the contractual and legal developments related to BIM keep up with the ever-increasing technological advances related to BIM.

This book deals with an overview of BIM (at Chapter 2); the BIM strategy (at Chapter 3); the BIM parties, teams and the information management process (at Chapter 4); the procurement phase, the planning phase, the production phase and the close-out phase of information management (at Chapter 5); the legal challenges raised by BIM (at Chapter 6); the contractual framework (at Chapter 7); collaboration and new contracts in respect of BIM (at Chapter 8); issues relating to design liability and insurance (at Chapter 9); information security matters (at Chapter 10); and dispute resolution involving BIM (at Chapter 11).

Peter Barnes
ISBN 978-1-83608-515-7
https://doi.org/10.1108/978-1-83608-512-620251002
Published under exclusive licence by Emerald Publishing Limited

Chapter 2
An overview of BIM

2.1. What is BIM?

BIM stands for building information modelling or building information management.

BIM is a digital process that relies on a comprehensive computerised virtual three-dimensional (3D) model of a built object (often referred to as a built asset) which is capable of reacting to changes in the 'virtual' world in the same way that the physical built asset would react when actually being constructed and/or when in operation. This virtual reaction is achieved by applying software applications to a computer model that are used to design, understand and demonstrate the key physical and functional characteristics of the built asset throughout its entire lifecycle (i.e. from its initial conception to its final demolition and/or disposal).

This computerised virtual 3D model also forms a shared knowledge resource of information about the built asset, providing data that can be collected, managed and used for a variety of reasons (again throughout the entire lifecycle of a built asset).

The BIM process creates data files of physical components and spaces to produce a sophisticated computerised model which contains the graphical representation of the built asset and, just as importantly, provides details of all associated information in the computerised model, including the built asset's geometry, spatial relationships, geographic information and various data attributes related to construction and operation, as data files that can be used and reused for other applications.

In many respects, BIM is closely linked to many other digital developments including smart devices, the Internet of Things, smart buildings and structures, smart cities and digital twins. These digital developments are briefly commented on below.

Smart devices are devices that are used to electronically monitor a wide range of parameters within a built asset, including things such as structural integrity, equipment performance, energy consumption, temperature, humidity and air quality. The data obtained from the various smart devices can be analysed in real time to identify and track patterns, trends and anomalies; this data can then be used to reduce energy costs and improve sustainability, among other things. These smart devices are often self-calibrating and, in those cases, in particular, the smart devices can detect or even resolve potential issues (including structural or performance-related) before they cause ongoing downtime, thereby helping to reduce maintenance costs and aiding with the improvement of productivity.

Linked to smart devices, the Internet of Things refers to a network of physical smart devices, vehicles, appliances and other physical objects that are embedded with sensors, software and network connectivity, allowing them to collect and share data. The Internet of Things is designed to enable different smart devices to communicate with each other and with other internet-enabled devices, thereby creating a network of interconnected devices that can exchange data and perform various tasks (often) autonomously. Through the linking and use of smart devices, the Internet of Things is advancing a new breed of smart buildings/smart built assets that are better aligned with the priorities of end-users and property managers. The Internet of Things facilitates operational systems that deliver more accurate and useful information for improving operations of the built asset, particularly in use.

Linked with the above concept of smart buildings/smart built assets are structures and built assets that utilise technology to monitor performance on a variety of metrics, such as comfort, efficiency and productivity, with the aim of optimising resource usage, efficiency and comfort for occupants and end-users. These buildings/built assets are equipped with a variety of modern and innovative technologies, from artificial intelligence (AI), virtual reality (which is a simulated experience that employs 3D near-eye displays and pose tracking to give the user the immersive feel of a virtual world) and augmented reality (which is technology that overlays video or images onto a display of the physical world to create an interaction between the end-user and the digital and the physical worlds).

The phrase 'digital twin' also relates to smart buildings/built assets. A digital twin is a virtual representation of an object or system designed to accurately reflect a physical counterpart (either one that has been constructed or one that has yet to be constructed). The digital twin spans the built asset's lifecycle and is constantly updated in real time using data from embedded sensors. By leveraging simulation, machine learning and reasoning, digital twins help make informed decisions and improve the original physical entity and serve various purposes, such as simulation, integration, testing, monitoring and maintenance. Essentially, digital twins allow an end-user to study and optimise real-world objects or systems in a virtual environment; this pairing of the virtual and physical worlds allows analysis of data and monitoring of systems to head off problems before they even occur, thus preventing downtime and helping to develop new opportunities and plans for future possibilities by using simulations of prospective actual events.

A development of the above, which largely concentrates on one built asset/facility, is the linking of several or many built assets/facilities so that, for example, the energy usage across a range of built assets/facilities is optimised; therefore, rather than having a smart building/asset, this can be developed into smart communities (sometimes referred to as smart cities). For example, if smart technology was used for the control of heating and hot water services, so that the required temperature in a built asset/facility was kept at an optimum level, there would only need to be a saving of a few percentage points to make enormous financial and energy savings across an entire community area (a smart city). As a further development of this, the optimum temperature level (for example) may itself be produced in part from algorithms that try to learn the end-users' behaviours and usage.

Of course, some of the above matters are being used in the present day, while others will come into use more and more in the coming years.

2.2. The levels of BIM and BIM maturity

Although becoming much less important in the current digital world and less often referred to and relied on, BIM was categorised in its earlier days by referring to BIM 'Levels' and BIM maturity. As the terms BIM Level and BIM maturity may still be referred to by others elsewhere and may be seen in other literature, a summary of those Levels and maturity is provided below.

Level 0 Level 0 of BIM relates to un-managed CAD in 2D, with paper or electronic data exchange, and simply uses 2D CAD files for design and production information. This is not really BIM as we now know it at all.

Level 1 Level 1 of BIM represents the first step towards genuine BIM and the use of 3D data to produce and present design. At this level of BIM, the designer is usually operating in isolation from other designers rather than working collaboratively with them. Designers at this level usually use managed CAD in 2D or 3D format. Normally, some standardised data structures and formats are used, and there may also be commercial data management provided by separate stand-alone finance and cost management software packages which are not integrated into the general BIM model.

Level 2 Level 2 of BIM is a managed 3D format held in separate BIM discipline software tools with data attached. This level of BIM introduces the first steps to utilising other digital add-ons, such as 4D software for construction time and sequencing data and/or 5D software for cost information.

Level 3 Level 3 of BIM (sometimes referred to as 'Open BIM') relates to a fully integrated and collaborative real-time project model. This level of BIM is capable of utilising 4D construction time and sequencing software, 5D cost information software, 6D facilities management information software and other software dimensions for other management information or functions.

It must be noted that despite the above, more often, at the present time, the concepts of BIM Levels and BIM maturity are incorporated into the overall concept of 'digital construction', which is a concept which recognises that too precise a definition and delineation of BIM Levels and maturity simply risks misunderstanding the BIM concept and may actually impede progress.

Of course, every project is different from another so the precise requirements in relation to the use of BIM for the various projects will also differ; therefore, it is now recognised more and more that rather than focusing on abstract terms such as BIM Levels and BIM maturity, it is more important to concentrate on BIM deliverables (i.e. what outputs are required from the BIM process) and to make sure that those BIM deliverables are clearly stated. For this to be effective, every party on a BIM project needs to know what it has to deliver and when, what it can expect the other parties to deliver and when, and the risks and obligations that surround this basic structure therefore need to be spelled out completely, clearly and unambiguously.

2.3. The benefits of BIM

The BIM process brings with it many benefits. These benefits relate not only to the design, engineering and construction phase of the process (sometimes referred to as the CapEX phase) but also, and perhaps even more so, to the operations phase of the built assets (sometimes referred to as the OpEx phase).

Some of the many benefits that are achieved through the use of BIM are set out below.

- The BIM process encourages collaboration of working and design between the various consultants and helps to streamline and coordinate communication and collaboration between teams by providing a centralised platform for sharing and updating project data in real time.
- The BIM process permits everyone involved in the process (including the client/end-user) to have a clear understanding of the concept design by providing the ability to visualise (e.g. by way of a virtual 3D model walkthrough) what is actually going to be built. This model walkthrough can also allow the built asset to be seen, virtually, at different times of day and from different angles. This possibility reduces the likelihood of misunderstandings at the very early design stages, reducing the prospect of disputes occurring at a later stage.
- Because of the collaboration noted above, the BIM process allows for a faster design and construction process.
- The BIM process allows for the full coordination of the various consultants' designs, thereby helping to prevent conflict between those designs.
- The BIM process provides the ability to identify physical collisions and clashes (e.g. identifying ductwork that would otherwise run into structural members) and as such can be rectified before construction starts on site, thus saving time and costs.
- The BIM process assists with better design and space planning which allows for the optimisation of the use of the built asset.
- The BIM process permits value engineering exercises to be carried out on all the works or on portions of the works, as applicable, projected on a virtual basis.
- By using the appropriate software, the BIM process can accurately and rapidly generate an array of essential estimating information, such as materials quantities and costs, size and area estimates, costs and productivity projections. This software approach can be carried out as changes to the scheme are made and/or variations are introduced, so that it is easy to see at a very early stage what the likely implications of the proposed changes will be, allowing a great saving on human resources.
- The BIM process allows for the modelling and remodelling of alternative layouts and options and for the comparison and selection of different construction techniques to see which options are most beneficial and cost effective.
- The BIM process helps to provide a very useful and successful marketing tool for all those involved in a project, particularly for consultants and contractors of a smaller size, as elements of the BIM process can help them compete in marketing terms with much larger organisations with much larger resources.
- The BIM process helps improve health and safety by predicting and foreseeing risks through advance visualisation of the construction process. Further, the use of 4D timelined models in design supports the principles relating to 'inherent safe design', 'safety by design' and the legislative duties on designers.
- The BIM process permits contractors to carry out a dry run of the proposed construction, making it simpler for them and their subcontractors to work out the safest

and most efficient way of carrying out the works in the virtual world before actually physically starting work on site.

- The BIM process assists with the ability to do more 'what if' scenarios, such as looking at various sequencing options, site logistics, hoisting alternatives and so on, which not only improves cost and time efficiencies but will inevitably result in safer working conditions.
- The BIM process facilitates fewer errors and corrections on site, leading to less rectification and remedial works, resulting in lower overall costs and greater productivity.
- The BIM process gives the opportunity for programme visualisation and makes it easier to understand the dependency that various activities have on one another. This ability is critical to the efficient and safe sequencing and execution of the works on site.
- The BIM process provides the ability to use information from the various computerised models to directly create fabrication drawings, thus avoiding what can often be a problematic and difficult step in the lead up to the construction process.
- The BIM process allows the opportunity for more prefabrication of components to be carried out off site, resulting in components that potentially will be at a higher quality and/or at a lower cost than by using the more traditional construction methods. This pre-fabrication approach should also have the effect of reducing waste.
- Through the use of specialist software, the BIM process assists in ensuring that a project complies with all of the required statutory and regulatory codes and requirements, as well as reducing any associated delays in respect of these matters.
- The BIM process helps maintain accurate project documentation, making it easier to track changes throughout the built asset's construction and its entire lifecycle.
- The BIM process makes it much simpler to produce 'as built' drawings (whether that be physical or electronic) and also the operation and maintenance manuals and other documents required on completion of the construction phase of the built asset.
- The BIM process allows for reduced capital expenditure and operational expenditure. The BIM process does this as it can provide a solution to the spiralling cost of energy for business and occupants alike by using sensors to focus energy usage only where it is needed and reducing consumption in areas that are not in use (even for specific times of the day/days of week etc.) and this approach can help save on unnecessary operational and energy costs. In addition, the insights derived from smart technology could lead to broader business decisions being made that have a positive impact on an end-user's profitability. For example, the process may identify opening times that align more closely to employee (and customer) working patterns and, simultaneously, help improve energy savings and reduce operating costs.
- The BIM process allows for evaluation during the early design phases, and also for updating thereafter, of the energy expected to be used over the lifecycle of the built asset, helping to improve the carbon footprint of the built asset and resulting in a 'greener' building. Also, the BIM process can be used to assess the environmental impact of a built asset, including energy consumption, water usage and carbon emissions, which all help in the design of more sustainable structures. Designing new built assets (and retrofitting existing ones) with smart building technology can also help dramatically reduce energy consumption, resource usage and, subsequently, reduce the built asset's overall carbon footprint. This of course applies to the carbon emissions

created in the construction process itself as well as to the energy used once a built asset is operational. It is important to remember that the effect on the environment needs to address the entire built asset lifecycle (from conception to disposal).

- The BIM process helps to create healthy buildings – for example, by using computer-driven fluid dynamics to assess indoor air quality, by performing analysis of discrete zones to assess the thermal comfort of individual occupants, by examining the feasibility and performance of different ventilation strategies, by managing moisture and condensation, by assessing reverberation and acoustic impacts and by measuring and designing ambient lighting.
- The BIM process can improve space utilisation. Alongside energy, expenses and resources, space is another key commodity that property owners and businesses value highly, and the BIM process can help facilitate the effective use of space. Everything from parking spaces to desks and meeting rooms can be tracked, organised and allocated through digital technology, ensuring that office and residential spaces provide the necessary access and resources for all their occupants.
- The BIM process provides an enormous amount of data generated by many participants relating to the design, construction and operation of the built asset which is integrated into the various BIM models and this permits the ongoing and improved analysis of data trends.
- The BIM process assists with maintenance, facilities management and lifecycle replacement issues. The data produced from the BIM process can be used to provide appointing parties/clients/end-users with a detailed digital representation of the building, making maintenance and operations more efficient through the use of predictive maintenance and real-time monitoring. If left undetected, building defects and faulty equipment often have a significant impact on safety, energy usage and wellbeing and, in the worst cases, can be catastrophic for both an organisation's finances and its reputation. The recent RAAC (reinforced autoclaved aerated concrete) crisis in the UK (which affected many public buildings including schools and hospitals) is a prime example of problems that can arise. The use of smart technologies can continuously monitor and collect data from each system within a built asset. They can also use sensors to identify potential structural weaknesses or faults and this helps to improve maintenance and aids in ensuring that any problems are dealt with quickly and (consequently) at a far lower cost.
- The BIM process can provide real-time insights and analysis. The continuous monitoring and collection of real-time data provides useful information and actionable insights for end-users, building managers or occupants to help them make informed decisions. These decisions may relate to the adjustment of energy usage and temperatures, lighting and indoor air quality. Smart building technology can help buildings become flexible and reactive to needs and demands, ensuring both optimised comfort and efficiency. They can also help facilities managers and teams make broader decisions based on workplace behaviour, improving the day-to-day running of the overall organisations concerned as well as the built asset itself.
- The BIM process allows for the automation and integration of the built asset with other systems. Integration and connection with systems within a built asset make automation a reality for building owners, managers and occupants. This possibility can have significant ramifications for building managers and facilities management teams, as

automating simple and repetitive processes can save them time and allow them to focus on higher-priority tasks and objectives. When these automated processes are based on real-time data, they also reduce the risk of human error arising.

- The BIM process allows for greater occupant comfort. A major selling point of the smart building design is how it can be used to create a comfortable working/living environment focused on wellbeing. In some cases, automation allows a building to respond to external weather conditions (possibly by the use of blinds or similar), ensuring, for example, the optimum internal temperature for its occupants. Lighting and air conditioning systems can also be adjusted to provide the highest level of comfort.
- The BIM process can help increase productivity. One of the very few positives of the COVID-19 pandemic was that it brought greater awareness to the general public regarding the importance of wellbeing and the relationship between health, wellbeing and productivity. For businesses in commercial buildings, smart technologies can be geared towards creating healthy environments and optimal working conditions, and this should result in more content and more productive employees.
- The BIM process can assist with enhancing physical security. A very important aspect of a built asset is its physical security. Smart building design can utilise technologies such as linked security cameras and sensors integrated with control and security systems to detect potential intruders or threats. In this way, a facility manager (or an end-user) can be alerted to any activity and, through the use of the above automation processes, will be able to respond easily and promptly.
- The BIM process can contribute towards an increase in value of the built asset. The growing popularity (and in many ways the now essential nature) of smart technologies makes any built asset equipped with them an attractive proposition to a purchaser or a renter. With the benefits of improved efficiency, greater occupant comfort, enhanced security and predictive maintenance, investors, owners and even occupants will usually pay more for properties and assets equipped with smart building solutions.

In addition to the more obvious benefits relating to the lifecycle of a built asset, as outlined above, BIM brings with it additional opportunities in the form of exploiting existing and new export markets for BIM 'know-how', BIM software and BIM-ready products.

The additional and collateral effect of the widespread adoption of BIM both domestically and internationally should bring with it a great opportunity for niche software developers and consultants, as well as other more construction-centric consultants who will have gained invaluable experience of the use of BIM.

Therefore, in overview terms, BIM potentially enables collaboration between clients (end-users), designers and contractors in technological ways (and possibly through contracts or other forms of agreement, which are dealt with in Chapters 7 and 8 of this book) that have not been available to the construction and civil engineering industries in the past.

While BIM is certainly not the solution to all problems, the BIM experience, digital simulations, rehearsals of all stages of the design, build and operate processes, and the information embedded within the BIM models, will facilitate well-informed decision making, which should result in better business outcomes, greater clarity, improved communication, less risk and, ultimately, better efficiency.

2.4. The challenges of BIM

Despite the large list of the potential benefits of the BIM process (as set out above), for the BIM process to continue to develop and grow, the professional advisers of the appointing party (normally the client or the end-user of a built asset) need to manage the appointing party's expectations about what BIM is, what it can do and what information it will produce (and at what cost).

For it to be fully effective, there is a need for more aware and actively involved appointing parties (clients/end-users). An appointing party who needs or wants a particular 'smart' built asset at the end of the process which satisfies its needs will need to be committed to the design process in the early stages in order to define what it requires from the built asset when it has been constructed. Obviously, everyone has to buy in at the beginning of the BIM process but, in this context, none more so than the appointing party.

Consequently, when BIM is introduced or specified for a built asset, there needs to be clarity and a common understanding of where to start, what to do and how to operate under a 'BIM project' as compared to a 'traditional project'. If there is inconsistency or uncertainty in this regard, this can cause confusion and divergence, leading to unnecessary barriers for the successful implementation of the BIM process.

Even though most informed commentary is that the facilities management sector has perhaps the most to gain from the use of the BIM process, presently, the facilities management sector appears to be the least involved and appears to have very little input into the BIM process, and this clearly needs to change. Therefore there needs to be an effort made at a very early stage of the design and construction phase to explain to the facilities management sector (and to the appointing parties/end-users) the benefits of BIM which will come into existence during the operational stage of a built asset.

There are obviously concerns regarding how to pass responsibility for the BIM model between the various parties participating in the BIM process. The design team typically contributes significant thought and creativity into the design process, and there are concerns that this input may largely be wasted as others (including possibly the contractor) assume responsibility for the design of the project, at which time the design may effectively be carried out again afresh some way into the design and/or construction process.

Consequently, contracts enabling collaboration and providing protection to participants in the BIM process need to be rapidly developed, and this matter is returned to later in Chapter 8 of this book.

Further, there is still considerable variety in the types of software that organisations use in the BIM process, and this leaves open the possibility of parties on a particular project using different software with consequent incompatibility and interoperability issues that may arise. Again, this is a matter that needs to be urgently addressed and, in reality, the solution is likely to be 'market-led'.

Collaboration between the participants involved in construction projects and in asset management is pivotal to the efficient delivery and operation of built assets. Organisations are increasingly working in new collaborative environments to achieve higher levels of quality and greater reuse of existing knowledge and experience. A significant outcome of these collaborative environments is the potential to communicate, reuse and share information efficiently, and to reduce the risk of loss, contradiction or misinterpretation.

Through collaborative working everyone involved in a project can easily and speedily share ideas, perceptions and interpretations. However, true collaborative working requires mutual understanding and trust and a deeper level of standardised process than has typically been experienced in the past. Current practices of silence, often for fear of liability or lack of knowledge, obviously need to be eliminated (when it is possible to do so), and a new process where decisions are made at the highest and most appropriate level of competency needs to be established to make full use of the various participants' knowledge. It is apparent that, unless properly addressed at an early stage, this issue, if left unaddressed, could be one of the greatest obstacles to the transformation and realisation of a fully optimised BIM project.

2.5. Abbreviations/acronyms/definitions/glossary of terms

In the section below an alphabetical list of the more common abbreviations, acronyms, definitions and other terms used in the BIM process are provided.

Appointed party (and lead appointed party)

An appointed party is an organisation (e.g. designers, contractors, subcontractors, suppliers etc.) that is responsible for creating and providing information, work or goods according to its appointment document.

An appointed party can be appointed by the appointing party directly or by a lead appointed party. An appointed party is a member of the project team, one or more delivery teams, and one or more task teams.

A lead appointed party is a party responsible for coordinating information between the delivery team and the appointing party. The lead appointed party is appointed by the appointing party and is a member of the project team and of one or more delivery teams.

The above relationships are set out in Figure 2.1.

Appointing party

The appointing party is usually the client, building owner or an end-user of the built asset in question.

Appointment

This is the agreement/contract that applies (at any level down the supply chain) to the works in question. This could, for example, be between the appointing party and the lead appointed party or between a lead appointed party and an appointed party. These appointment connections are set out in Figure 2.2.

Figure 2.1 Relationship between the appointing party, a lead appointed party and the appointed parties

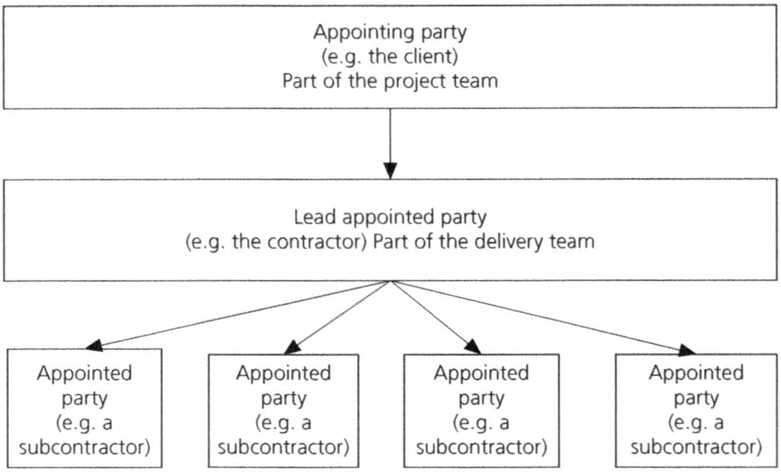

Figure 2.2 Appointment connections

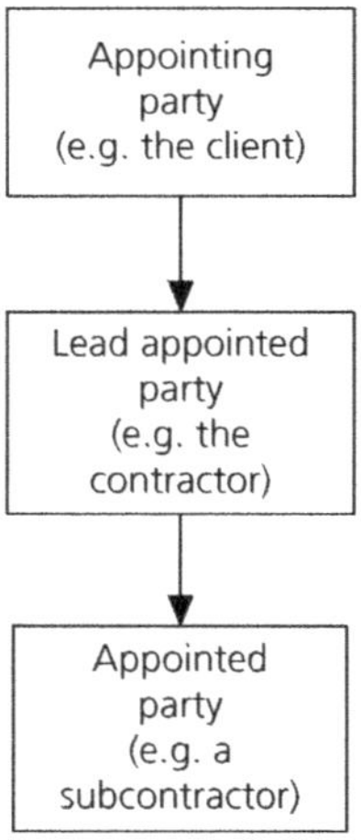

Asset information model (AIM)

The asset information model is developed from the project information model and contains information related to the use, management, maintenance and operation of the built asset. The asset information model is to be revised as and when the built asset changes in the future. The asset information model is normally referred to as the AIM. The relationship between the project information model and the asset information model is set out in Figure 2.3.

Figure 2.3 Relationship between the project information model and the asset information model

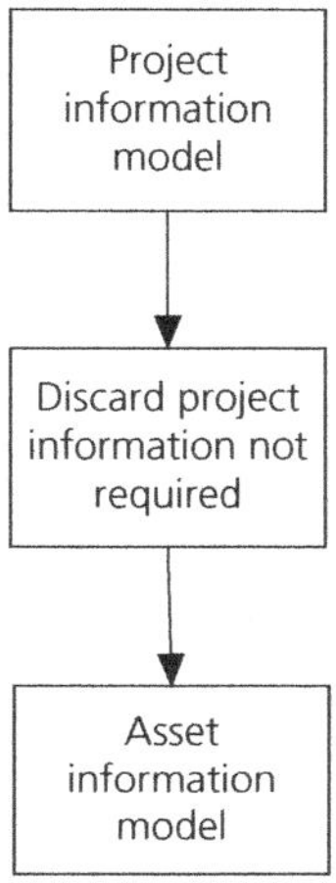

Asset information requirements (AIR)

The asset information requirements set out the information needed by the client/end-user for the ongoing operation and maintenance of a built asset. The asset information requirements are normally referred to as the AIR.

BIM coordinator

The BIM coordinator essentially has a design role with responsibility for model coordination and clash detection. The BIM coordinator does not normally carry out any design but is only concerned with model coordination and clash detection. This BIM coordinator role is set out in Figure 2.4.

BIM execution plan (BEP) (pre-contract and post-contract)

A BIM execution plan is a document that outlines how the BIM process will be utilised and applied on specific projects. The BIM execution plan is normally referred to as the BEP. The BEP defines project objectives, the scope of BIM use, roles and responsibilities, information exchange protocols, software standards and project-specific requirements (including model requirements). It explains how the information management aspects of the works to which it relates will be carried out by the relevant delivery team as may be identified in the information particulars and/or provided in accordance with the information standards required.

BIM information manager

The BIM information manager is essentially a procedural gatekeeper, policing the model(s), the common data environment (CDE) and the associated BIM and information processes to ensure that they follow the requirements of the respective appointments and that the data that is held is accessible but secure.

Figure 2.4 The BIM coordinator's role in model production

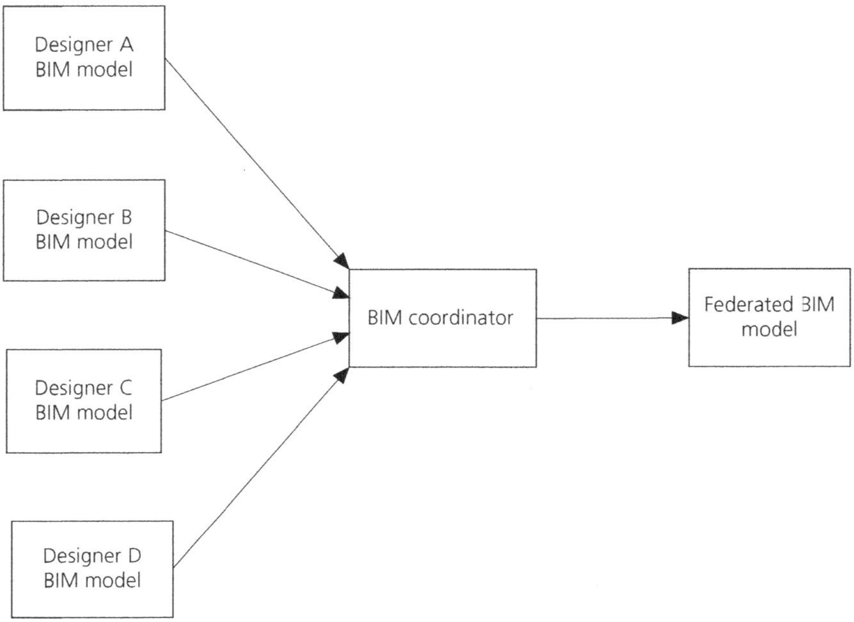

BIM protocol

The BIM protocol is now referred to as the information protocol which is defined below.

COBie (Construction operations building information exchange)

Construction operations building information exchange (COBie) is a spreadsheet data format (sometimes simply in an Excel format) which contains digital information about a built asset. COBie is often thought of as a subset of a building model.

Common data environment (CDE)

The common data environment (CDE) is a central electronic-based repository to collect, manage, store, archive and disseminate information developed during the BIM process.

Construction Playbook

The *Construction Playbook* (HMG, 2020) is a UK Government document that outlines key policies and guidance for assessing, procuring and delivering public works projects and programmes in the UK. Specifically, it reinforces the importance of the BIM process and the digital requirements for public sector clients and contractors. The *Construction Playbook* aims to accelerate the benefits of BIM implementation and digital transformation in construction by synthesising experiences from around the world. It emphasises continuous improvement in safety, cost, speed, quality, data sharing, transparency and investment. The *Construction Playbook* seeks to deliver the ambitions set out in *Transforming Infrastructure Performance:*

Roadmap to 2030 (Infrastructure and Projects Authority, 2021) and brings together best practice from across the public sector. The *Construction Playbook* (HMG, 2020) says that if its policies are adopted, this will

- set clear and appropriate outcome-based specifications that are designed with the input of industry to ensure that continuous improvement and innovation are achieved
- favour longer-term contracting across portfolios, where it is appropriate
- standardise designs, components and interfaces as much as possible to improve quality, safety and performance and to reduce the environmental impact
- drive innovation and modern methods of construction through standardisation and aggregation of demand, increased client capability and setting clear requirements of suppliers
- create sustainable, win-win contracting arrangements that incentivise better social, economic and environmental outcomes, improve risk management and promote the general financial health of the sector
- embed social value through improved quality, safety and performance and reduced environmental impact at the heart of programme delivery
- strengthen financial assessment of suppliers
- increase the speed of end-to-end project and programme delivery by investing up front with time and resources to set projects up for success.

Data drops (information milestone)

Data drops are the times when collected data is accumulated and/or reported on. Data drops are also sometimes referred to as 'information milestones'.

Delivery team

This is the team put together to deliver some particular aspect of the BIM deliverables. The applicable lead appointed party and the applicable appointed parties are part of the delivery team. This team structure is shown in Figure 2.5.

Figure 2.5 The delivery team's structure

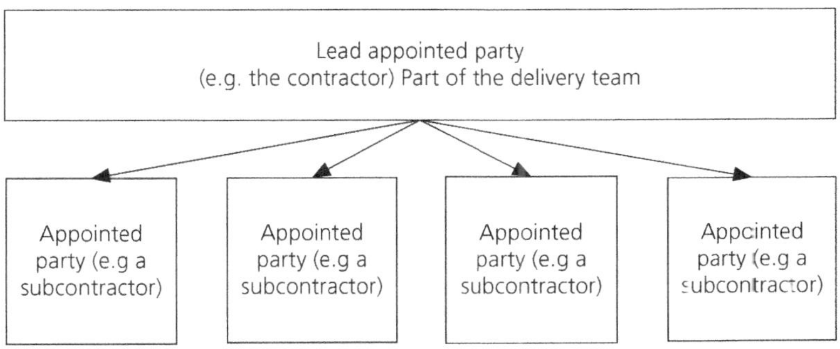

Deliverables (BIM deliverables)

BIM deliverables are the outputs that the appointing party requires as a result of the BIM process.

Exchange information requirements (EIR)

The exchange information requirements (normally simply referred to as the EIR) are detailed appointment-specific information requirements that precisely specify what information is needed at the time of each data drop (information milestone) to enable the various BIM participants to complete the necessary information activities by the required time during the delivery and operation phases of a built asset.

Federated information model (FIM)

The federated information model is an information model consisting of and created from connected but distinct individual project information models. The federated information model is normally referred to as the FIM. This development is shown in Figure 2.6.

Information particulars

The information particulars are the specific details within the information protocol (previously known as the BIM protocol) which set out the BIM-specific requirements for a project, including, among other things, information regarding the exchange information requirements, the BIM execution plan, the master information delivery plan, the task information delivery plan, the risk register and the responsibility matrix.

Information protocol

The information protocol (previously known as the BIM protocol) is a document which defines the various parties' and/or participants' rights, roles and duties in compliance with the project's required BIM standards and procedures. A project-specific information protocol is required to help ensure that all project stakeholders, including end-users, architects, engineers, contractors, subcontractors and facility managers, work together efficiently and effectively to leverage the full potential of the BIM process to deliver better-designed, better-constructed and better-managed built assets. The lead appointed party's own appointment document will

Figure 2.6 Federated information model developed from project information models

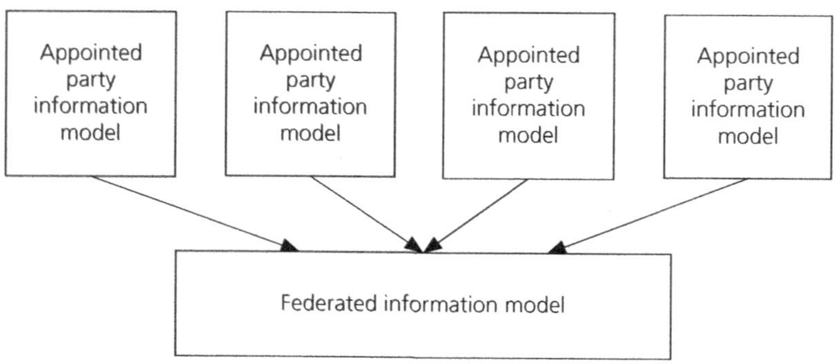

(normally) contain the project's information protocol, and the project's information protocol document will also be included in all the appointed parties' appointment documents (with any appropriate and necessary changes made to reflect the requirements of each separate appointment but ensuring a good degree of consistency across the various appointment documents).

Information standard

The information standard is the BIM standard that applies in respect of a particular project, normally based on the BS EN ISO 19650 (BSI, 2019) series of international standards, as may be amended or superseded from time to time.

Level of detail (LOD)

The level of detail (LOD) defines the level of complexity required for various components or elements within the various BIM models at different stages of the project.

Level of geometry (LOG)

The level of geometry (LOG) defines the level of geometry (i.e. a display that can vary from symbols or coarse concept bodies to a detailed representation) required within the various BIM models at different stages of the project.

Level of information need (LOIN)

The level of information need (LOIN) is the framework which defines the extent and granularity (i.e. refinement or coarseness, as applicable) of information as may be identified and/or required. Part of the principles of the BIM process is that neither too little nor too much information should be produced.

Master information delivery plan (MIDP)

The master information delivery plan (MIDP) sets out when project information is to be prepared across a project, who is responsible for preparing that information, and what protocols and procedures will be used to develop the required information. The summary of the information as set out in the master information delivery plan is collected from a series of individual task information delivery plans. The creation of the master information delivery plan is shown in Figure 2.7.

Figure 2.7 Master information delivery plan developed from the task information delivery plans

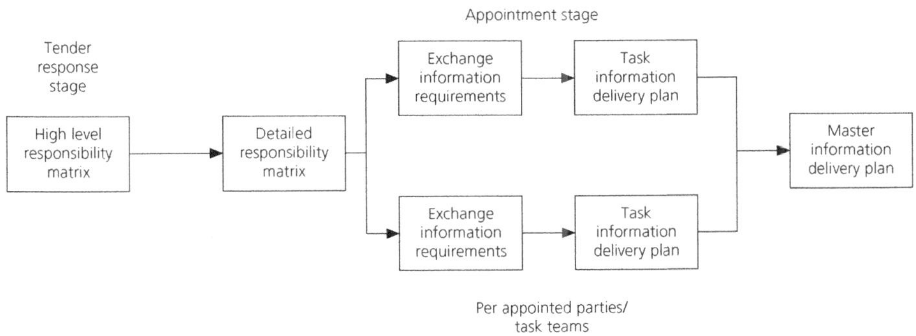

Organisational information requirements (OIR)

The organisational information requirements set out the organisational objectives, needs and necessities of the client/end-user in respect of the information requirements related to the built asset in question. The organisational information requirements are normally referred to as the OIR.

Project implementation plan (PIP)

The project implementation plan sets out the capability, competence and experience of potential lead appointed parties and appointed parties bidding for work on a built asset operating under the BIM process and details the required quality documentation provided. The project implementation plan is normally referred to as the PIP.

Project information model (PIM)

The project information model is developed during the design, production and construction phase (the delivery phase) of a project and consists of documentation and graphical and non-graphical information defining the delivered built asset. The project information model is normally referred to as the PIM.

Project information requirements (PIR)

The project information requirements set out information needs for the delivery phase of a built asset. The project information requirements are normally referred to as the PIR.

Project team

The project team consists of all participants (including the appointing party, the lead appointed parties and the appointed parties) involved in the delivery process of a built asset. This team structure is shown in Figure 2.8.

Responsibility matrix (general and design)

The responsibility matrix identifies which participants are responsible for providing what information and/or design and when.

Risk register

The risk register identifies all risks associated with the prospective delivery team's management and production of information throughout their respective appointments, as well as showing how they intend to manage those risks.

Security management plan

The security management plan is the document (or part of a document) identified in the information particulars setting out the security requirements for carrying out the works to the built asset in a security-minded way, including any policies, protocols, processes and procedures related to the matter of security.

Task information delivery plan (TIDP)

The task information delivery plan shows who is responsible for each element of the information that is deliverable, in what format that information is to be provided and at what date(s) the information is to be provided. The task information delivery plan is normally referred to as the TIDP.

Figure 2.8 The project team's structure

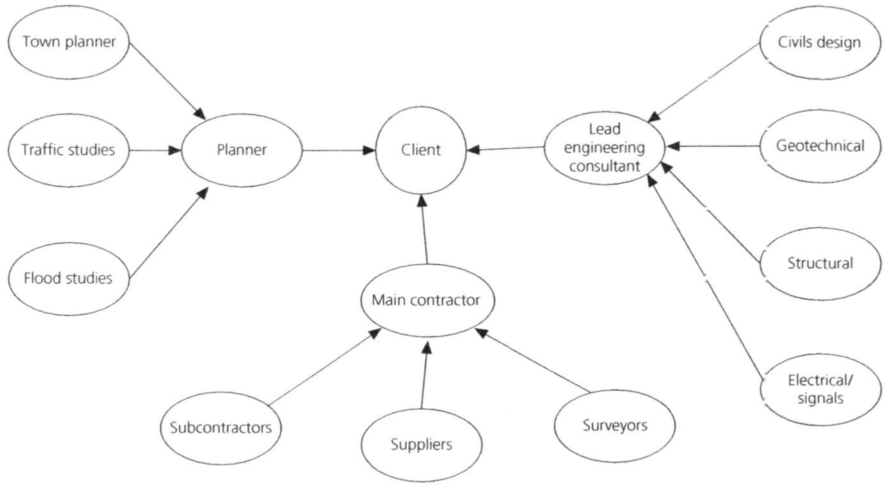

Task team

The task team comprises only those appointed parties (within a delivery team) (directed by a lead appointed party) tasked with performing a specific task in respect of the BIM process. There are normally many task teams.

2.6. Summary

- BIM stands for building information modelling or building information management.
- BIM is a digital process that relies on a comprehensive computerised virtual 3D model of a built asset.
- This computerised virtual 3D model forms a shared knowledge resource for information about the built asset, providing data that can be collected, managed and used for a variety of reasons.
- In many respects, BIM is closely linked to other digital developments such as smart devices, the Internet of Things, smart buildings and structures, smart cities and digital twins.
- In the past, but not so much now, the maturity of BIM was referred to by the term BIM 'Levels', with BIM Level 0 to BIM Level 4 applying.
- More often, presently, the concept of BIM Levels and BIM maturity are incorporated into the concept of 'digital construction'.
- The BIM process brings with it many benefits. These benefits relate not only to the construction phase of the process but also to the operations phase of the process.

- Despite the large number of benefits of BIM, challenges continue to exist and expectations of what BIM can presently achieve need to be managed.
- Collaboration between the participants involved in BIM is pivotal to the efficient delivery and operation of built assets through BIM.
- BIM is a process that relies on many acronyms and defined terms, and an understanding of these acronyms and definitions aids the understanding of the BIM process.

emerald PUBLISHING ice Publishing

Peter Barnes
ISBN 978-1-83608-515-7
https://doi.org/10.1108/978-1-83608-512-620251003
Published under exclusive licence by Emerald Publishing Limited

Chapter 3
The BIM strategy

3.1. The BIM strategy

BIM is not only about software or computer modelling for construction purposes; it is also about obtaining the right information and data at the right time to allow better decisions to be made more efficiently about a built asset during its construction and over its entire lifecycle.

In respect of information and data gathering, the purpose of a BIM strategy is for the appointing party (i.e. normally the entity responsible for initiating the project (sometimes referred to as the client)) to establish what it wants or needs to obtain in respect of information and/or data (this information and/or data being known as the BIM deliverables) through the use of BIM for a particular built asset or a series of built assets, and how and when it will implement the resulting BIM deliverables, taking into account the supporting processes, technology, tools and people that will be required to action the BIM strategy decided on.

As part of the process of arriving at a BIM strategy, a statement of need is normally produced first by the appointing party in an attempt to define and describe the outline BIM requirements and deliverables that the appointing party is looking to obtain. A BIM strategic brief is then developed from that statement of need, which is intended to describe the appointing party's BIM requirements in sufficient detail to allow for (at the very least) the appointment of the required consultants at the first stage of the BIM process.

In reality, a clearly defined BIM strategy is at the heart of the successful implementation of BIM in respect of built assets, and that is why sufficient time should be spent in obtaining an accurate and properly detailed BIM strategy from a clearly thought-out statement of need, at the outset. It is only when an appointing party is clear about what it is hoping to achieve from BIM that it can start to contemplate what a successful BIM outcome would look like, and it is only then that it can consider what will be needed to make that successful BIM outcome happen.

The BIM strategy will almost certainly be unique and distinct to each appointing party and to each built asset as it must reflect that appointing party's own ambitions and requirements from BIM generally as well as for the particular built asset or built assets in question.

If an independent BIM adviser is appointed in the early stages of a project, this can help an appointing party to assess its long-term business objectives and to define how BIM could be used by and be useful to it. The appointed independent BIM adviser can also help the appointing party make the most innovative use of BIM during the entire lifecycle of a built asset in an effort to reduce costs, increase operational efficiency and improve business outcomes.

To be of the greatest value to an appointing party, the independent BIM adviser needs to

- be able to interpret the appointing party's business and organisational needs
- understand the whole project development process, from inception to operation and flowing through to facilities management
- understand the interfaces between project management and facilities management
- have a clear understanding that BIM is much more than just creating a computerised BIM model
- be able to advise on process improvements and technological solutions
- be able to advise on the most effective capture and use of digital data and information
- be able to advise on assessing the competence and capability of potential lead appointed parties and (where applicable) appointed parties
- be able to assess education and training needs in respect of the BIM process.

Advice from a BIM adviser is particularly important in the very early stages of a project as this is normally when an appointing party needs guidance on understanding what can be achieved from the BIM process and needs assistance in preparing its statement of need, the project BIM brief and (flowing from those documents) the exchange information requirements for the particular appointing party and built asset in question.

3.2. The project BIM brief

Whereas the general project specification defines the nature of the built asset that the appointing party wishes to procure and have built (i.e. the construction aspects of the built asset), the project BIM brief is a document that defines the appointing party's requirements and expectations for a project specifically in respect of the BIM process.

Consequently, the project BIM brief identifies goals, objectives and the various participants' responsibilities in respect of the BIM process and needs to be developed in the pre-design phase of the proposed built asset so that it is fully effective throughout the entire BIM process (from commencement to completion).

The project BIM brief is developed from the BIM strategy as referred to above and often includes the following items

- a description of the proposed built asset
- a description of the appointing party
- a description of the appointing party's brand, culture and organisation
- a description of the appointing party's vision, mission and objectives
- a list of the outcomes (BIM deliverables) that the appointing party expects from the BIM process
- a description of the appointing party's priorities and the criteria that will be used to measure success arising from the BIM process

- the appointing party's organisational structure and decision-making processes
- the changes/benefits to the appointing party that the project will bring about
- details of any interfaces in respect of the BIM process with other projects
- the appointing party's policies that may be applicable to the project/built asset (e.g. transport policy, energy policy, natural ventilation policy, sustainability policy etc.)
- the appointing party's preferences for the project (e.g. image, use of local materials, use of landscape etc.) and quality expectations (including health and safety, sustainability and design quality)
- a list of specific BIM deliverables required from the various participants (e.g. the design team, the contractor and the facilities management provider)
- a description of the principles that will be adopted in the development of the design
- an explanation of the minimum BIM modelling requirements
- the budget constraints applicable
- the project programme and any key milestones required.

While the project BIM brief may often start off in rather outline terms, it will almost invariably become increasingly detailed throughout the project concept and detailed design stages and is likely, ultimately, to include very specific and detailed information.

Even though the project BIM brief may be a continually developing document, the general view is that the project BIM brief should be frozen at the end of the concept design stage; change control procedures should then be introduced from that point on to prevent further changes being made without appropriate justification and authorisation being made by the appointing party.

The project BIM brief is likely to be presented in a simple report format; however, where possible, information and requirements should be scheduled in a database or spreadsheet format that will be easier to expand in the future and which will be easier to use to test whether the BIM proposals satisfy the BIM requirements later in the project.

An example of the contents of a project BIM brief is set out below.

1. Contents
2. Project information
3. Project schedule
4. Key project contacts
5. Project goals
6. BIM use competency requirements
7. Appointee-specific requirements
8. Project deliverables
9. Reference documents and standards.

3.3. Exchange information requirements (EIR)

The exchange information requirements (EIR) were previously known as the employer's information requirements but, in recognition of the fact that the document may not *only* set out the employer's information requirements, and also to be in line with the latest BIM standards (as dealt with later in this chapter), the document is now called the exchange information requirements.

A comprehensive and properly managed set of exchange information requirements is fundamental to successful information management on a BIM project; the exchange information requirements provide the overall framework to be used by each and every delivery team member working on a BIM project.

The EIR defines the information about the built asset that the appointing party wishes to procure to ensure that the design is developed in accordance with its needs and which enables it to operate the completed development of the built asset effectively and efficiently.

The information in the EIR will be detailed project-specific requirements that specify precisely what information and data need to be delivered (this information and data normally being referred to as the BIM deliverables), when that information and data need to be delivered, and sets out the standards and processes to be adopted at each required information milestone (an information milestone usually being referred to as a data drop).

The EIR is a specification detailing the information required by the appointing party for all information exchanges with a lead appointed party, an appointed party and a delivery team (as dealt with in more detail in Chapter 4 of this book).

The appointing party establishes the EIR for a lead appointed party, and then, based on that document, with amendments made as necessary and as required, the lead appointed party in question sets its own EIR for each of the separate appointed parties under its control, thereby formulating relevant and consistent information requirements down to each and every appointed party.

An EIR document is normally divided into three parts, as follows

- technical
- management
- commercial.

Each part consists of typical contents, as summarised below.

Technical

This part of the EIR describes the technical information requirements in respect of BIM and includes

- the exchange formats – i.e. the file formats and versions with which the participants are to carry out data exchanges, so that everything takes place in a format that can be used by and is accessible to the appointing party

- the level of detail – in line with the design needs, the minimum levels of detail, geometry, information need and model definition to be achieved for each model and for each technical element of each model must be defined.

The level of detail defines the level of complexity required for various components or elements within the BIM model at different stages of the project.

The level of geometry controls the level of detail of the geometry in the BIM models. The required display can vary from symbols or coarse concept bodies to a detailed representation with (for example) manufacturer-specific features. There are various commonly used geometric levels used when referring to the level of geometry from the lowest level, LOG 100 (usually used in a mass model, at the concept planning stage, which basically just indicates that a certain object has been designed for that particular location) to the highest level, LOG 400 (where all the real characteristics of the model element are included in the design, together with the exact quantities and dimensions, as well as information on production, assembly and installation).

The level of information need is the framework which defines the extent and granularity (i.e. refinement or coarseness, as applicable) of information and data as may be required. The purpose of the level of information need is principally to ensure that there is no wastage in providing information and data in excess of that which is actually needed (this requirement for enough information but not more information than is required is a basic tenet of the BIM process).

The level of definition is simply the aggregate of the level of detail and the level of information need.

- Software platforms – the software infrastructure and collaborative platform to be used must be stated and must be made clear.
- File size and format – the size of the model files, the size of extranet uploads or emails and the file formats that can define the size of volume must be stated.
- Coordinates – a common system of coordinates must be identified because all project models need to have the same points of coordination so that they are compatible with one another.
- Level of technical training requirements – the appointing party must set out the BIM training that is required to be carried out to enable a prospective participant to be able to take part in the proposed BIM process.

Management

This part of the EIR outlines the project process management requirements.

- The standards and procedures to be followed must be stated to ensure the correct flow of information, data exchanges, delivery and process security.

- The required roles and responsibilities of the various participants in the BIM process must be clearly set out.
- The information exchange dates relative to the delivery milestones (the data drops) need to be provided to ensure that information is delivered at the required time.
- The required data security measures to protect sensitive data must be stated.
- The coordination process that will be used, which includes methods of control and interferences, tolerances, resolution and so on, must be provided.
- The collaborative methods and procedures that the lead appointed parties, the appointed parties and the various teams must follow must be set out.
- A schedule of model review meetings that are needed to control the programme and to provide assistance to the various delivery teams during the performance of the appointment(s) must be provided.
- The design, health and safety management requirements – that is, the design requirements related to site health and safety that need to be put in place – must be defined.
- Any applicable constraints and limitations for communications under the appointment(s) must be stated.
- A compliance plan must be provided, setting out the details of the methods to be used to ensure the required quality of the produced information and data.
- The requirements for an asset information delivery model – that is, a definition of the minimum standards and a data scheme through which the asset information can be managed – must be stated.

Commercial

In this part of the EIR, the requirements related to the project goals and objectives are listed, including

- the strategic objectives – that is, the purpose of the BIM information requirements
- the BIM results to be achieved (i.e. the BIM deliverables). This part must define the BIM results to be achieved in detail, in terms of modelling, computerisation and coordination
- a skills assessment checklist, through which skills and competences of bidders for the project works in respect of the management and exchange of information can be analysed
- an initial responsibility matrix setting out the various participants' responsibilities for the model or information production in line with the defined project stages.

An example of the contents of an EIR document is set out below.

1. Introduction
    1.1 Exchange information requirements purpose
    1.2 Information management goals and objectives
    1.3 Information requirements hierarchy and progression

2. Information requirements
    2.1 Information purpose
    2.2 Plan of work
    2.3 Information delivery milestones
    2.4 Information security requirements
    2.5 Spatial coordination requirements
    2.6 Project information model
    2.7 Asset information model
    2.8 Information management key performance indicators
    2.9 Health and safety and design construction risk management
3. Information standards, information production methods and procedures
    3.1 Information standards
    3.2 Information production methods and procedures
4. Tables
    4.1 Information delivery milestones
    4.2 Spatial coordination information requirements
    4.3 Key performance indicators
    4.4 Health and safety information requirements
    4.5 Required exchange formats

3.4. BIM standards

To encourage standardisation and collaboration, and by focusing on the production, exchange and use of information and data in the BIM process as a means of delivering improved performance across the whole lifecycle of a built asset, it is important that standardised BIM specifications and standards are established and used.

In the past in the UK, British Standards and/or 'Publicly Available Specifications' (PAS) were often used for this purpose but, in respect of the current BIM process, these documents have now largely been replaced by the international BS EN ISO (the 'International Organization for Standardization') standards. In particular, the BS EN ISO 19650 (BSI, 2019) standard is an international set of standards of good practice for managing information over the whole lifecycle of a built asset using BIM.

BIM standards play a crucial role in standardising BIM practices, and this facilitates collaboration and standardisation among project stakeholders, which improves information management and thereby ensures that there is an efficient use of BIM throughout the lifecycle of built assets.

It is of course very important to bear in mind that current recognised international BIM standards are likely to evolve and/or change over time, so it is essential that the latest versions and/or updates and/or amendments of the appropriate BIM standards are identified and are used at the appropriate times.

With the above caveat very much in mind, the following BS EN ISO 19650 (BSI, 2019) standards presently apply to the BIM process in the UK.

BS EN ISO 19650-1

BS EN ISO 19650-1 (BSI, 2018a) provides an introduction into the concepts and principles of the information management process, and this is expanded on in the following parts of BS EN ISO 19650.

BS EN ISO 19650-2

BS EN ISO 19650-2 (BSI, 2018b) details the process for the management and collaborative production of information during the delivery phase of built assets and deals principally with the exchange information requirements, the pre-contract BIM execution plan (which is intended to address the matters raised in the exchange information requirements), the project implementation plan and a master information delivery plan, which sets out when project information is to be prepared across a project, who is responsible for preparing that information and what protocols and procedures will be used to develop the required information. The summary of the information as set out in the master information delivery plan is derived from a series of individual and separate task information delivery plans which show who is responsible for each element of the information that is deliverable, in what format that information is to be provided, and at what date the information is to be provided.

BS EN ISO 19650-3

BS EN ISO 19650-3 (BSI, 2020a) sets out the information management process for the management and collaborative production of information during the operational phase of assets. This also covers the transfer processes to create an asset information model (for use in the operational phase of the asset) from the project information model (used during the design and construction phase of a built asset). The asset information model will obviously need to be revised as and when the built asset changes in use.

BS EN ISO 19650-4

BS EN ISO 19650-4 (BSI, 2022a) relates to the recommended concepts and principles for the exchange of information between parties throughout the lifecycle of an asset.

BS EN ISO 19650-5

BS EN ISO 19650-5 (BSI, 2020b) provides a framework to assist organisations in adopting a security-minded approach to managing information relating to sensitive assets for projects employing BIM and for any built asset where asset information is created, stored, processed and viewed in digital form.

BS EN ISO 19650-6

BS EN ISO 19650-6 (BSI, 2023) relates to the requirement to identify, record, use and share information on health and safety risks which may result in harm to any person involved in a built asset from its outset and through to the end of its lifecycle.

3.5. BIM model requirements

At an early stage of the BIM process, it needs to be determined (initially and principally by the appointing party) what BIM models are required.

By way of an introductory comment to this section, a BIM model is a computerised 3D model-based process that allows multiple construction professionals to collaborate on the planning, design, construction and subsequent use of a built asset.

It must be remembered that a BIM model is not just a 3D drawing but is a digital representation of the physical and functional characteristics of a built asset and as such is a shared knowledge resource for information and decision making throughout the lifecycle of a built asset.

The extended use of 3D intelligent design models has led to reference to terms such as 4D model (i.e. adding time dimension software to the model) and 5D (i.e. adding quantities and cost of materials software to the model), as well as others that may be referred to as 'XD' dimensions for BIM models. It is often simpler to think of the 3D base computer model as a 'platform' onto which other applications (e.g. time, cost, environmental matters etc.) are built that may be used through the planning, design, construction and facility operation processes of the built asset. These further applications can deal with various types of information and data that are required in respect of a particular built asset; clearly the extent of these applications can be almost limitless, depending on the requirements of the appointing party. The formation of BIM models is shown in Figure 3.1.

In reality, there will rarely be a single BIM model on a typical project. For example, the design team may have a general architectural design model, the structural engineer may have a structural design model, the contractor may have a construction model, specialists may have fabrication models, and there may be (and often are) many other models in use also. The aim of the BIM process is to ensure that all the various models communicate seamlessly with each other; however, under the current technology (and software) limitations, that is not always possible and that is an issue that needs to be addressed going forward.

Figure 3.1 Formation of various BIM models from the 3D model platform

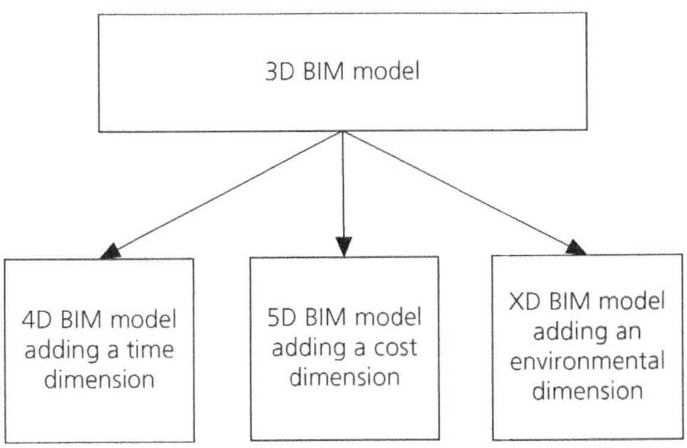

Therefore, in summary, a BIM model is a repository for both construction details and for information and data storage (in conjunction with the common data environment – dealt with in Chapter 5 of this book).

In the past, computerised models were largely static and, even assuming that the design professional could recognise when information needed to be preserved, it is unclear how that would have been accomplished in practice. In contrast to this, the more advanced modern dynamic computerised models can and do evolve, often on a self-automated basis, and this ability inherently and automatically involves the replacement of older information with newer information, and the overwriting of the older model or the discarding of obsolete data in the older model. It is important to note that while such a system can track revisions, unless the necessary software is in place they may not be able to accurately roll back every change made to the model, and this is something that must be kept in mind. As we will see later in this book, the ability to roll back time and freeze activities at a particular time could be important when disputes are considered.

Project information model (PIM)

The project information model is a computerised model that is developed during the design, production and construction phase (the delivery phase) of a project, and consists of documentation, graphical and non-graphical information defining the delivered (or to be delivered) built asset.

The project information model sits in the project's common data environment (dealt with in Chapter 5 of this book) and the collection of this information and data corresponds to the requirements as set out by the appointing party in the exchange information requirements.

The project information model consists of many different data storage areas (normally referred to as 'containers'), models, databases, spreadsheets and other documentation, all of which augment gradually together with the development of the delivery process.

In the initial stages, the project information model may comprise of only the scheme architectural model and some spreadsheets with basic information and data requirements included but, as the design and construction phase further develops, the project information model is likely to contain a vast amount of further design and construction information and data that is accumulated during the progress of the works.

The 3D model

The 3D model, among other things, provides for

- model walkthroughs
- clash detection
- project visualisation
- virtual mock-up models
- prefabrication.

The level of construction information in a BIM 3D model means that prefabrication can be utilised and there is a greater assurance that prefabricated components will fit once on site.

As a result, more construction work can be performed off site, cost effectively, in controlled factory conditions, and then efficiently installed on site.

The 4D model

The 4D (time) model provides for the following.

- Construction planning and management aspects – a 4D BIM model provides a means of verifying site logistics and operations by including software tools to visually depict the space utilisation of the site throughout the construction of a project. This type of model can include temporary components such as cranes, lorries and fencing. Traffic access routes for lorries, cranes, lifts and other large items can also be incorporated in the model as part of a site logistics plan. Software tools can further be used to enhance the planning and monitoring of health and safety precautions needed on site as the project progresses.
- Schedule visualisation – by watching the schedule visualisation, project members will be able to make sound decisions based on multiple sources of accurate real-time information. Within the BIM model, a chart (or similar) can be used to show the critical path and to visually show the dependency of some activities on other activities. As the design is changed, the more advanced BIM models will be able to automatically identify those changes that will affect the critical path and indicate what their corresponding impact will be on the overall delivery date of the built asset.

The 5D model

The 5D (cost) model provides for the following.

- Quantity take-offs – the model includes software that allows for the accurate and rapid generation of an array of essential estimating information, such as materials quantities and costs, size and area estimates, and productivity projections. In more advanced models, as changes/variations are made, the estimating/real-time cost information automatically adjusts.
- 'Real-time' cost estimating – cost data can be added to each object, enabling the model automatically to calculate an approximate estimate (or an accurate assessment) of material costs.
- Whole-life cost and lifecycle cost information – whole-life costing is a methodology for the systematic economic consideration of all whole-life costs and benefits over a period of analysis. Lifecycle cost is the cost of a built asset, or its part, throughout its lifecycle while fulfilling the performance requirements. Broadly speaking, lifecycle costs are those costs associated directly with constructing and operating the built asset, while whole-life costs include other costs such as land, income from the building and support costs associated with the activities within the building.

The 6D model

The 6D (facilities management) model provides for the following.

- Improved space management – the room and area information in BIM models are the foundation for good space management and the software in a facilities management model can help determine optimum room sizes and other spaces.

- Streamlined maintenance – the information about building equipment stored in BIM models can eliminate many days/weeks of effort to accurately populate maintenance systems.
- Efficient use of energy – the software in a 6D BIM model can help facilitate the analysis and comparison of various energy alternatives to help facility managers dramatically reduce environmental impacts and operating costs.
- Economical renovations – better information about existing conditions can help reduce the cost and complexity of carrying out building renovation projects.
- The facilities management model with the required software installed can assist with the lifecycle management of a built asset.

The 7D model

The 7D ('green') model provides for the following.

- Energy analysis – in a parametric building model, much of the data needed for supporting design analysis is captured naturally as the design of the project proceeds. The model contains the necessary level of detail and reliability for the imbedded software to complete the required energy analyses and makes it possible for routine analyses to be carried out directly by designers for their own baseline energy analyses, thus providing immediate feedback on design alternatives early on in the design process.
- Lifecycle analysis – an assessment of the environmental impact of a built asset throughout its lifecycle, using some form of environmental profile software methodology.

Federated information model (FIM)

A federated information model is an information model consisting of connected, but separate and distinct, individual information models.

Asset information model (AIM)

The asset information model is developed from the project information model and contains information related to the use, management, maintenance and operation of the built asset. The asset information model will itself need to be revised as and when the built asset changes. At the end of the project, some elements of the project information model are transferred into the asset information model and unnecessary data (normally referred to as residual project information) is archived. The development of the asset information model from the project information model is shown in Figure 3.2.

The asset information model normally comprises graphical models, non-graphical data and all necessary maintenance, operational and management documentation for objects in the model and the built asset. The appointing party, end-users and facility managers need to ensure that the asset information model is created before handover of the built asset and the end of the construction phase so that it can be used during the operational phase of the built asset.

The creation and the structure of information and data in an asset information model should be such as to enable transfer of that information and data to the appropriate computer-aided

Figure 3.2 The development of the asset information model from the project information model

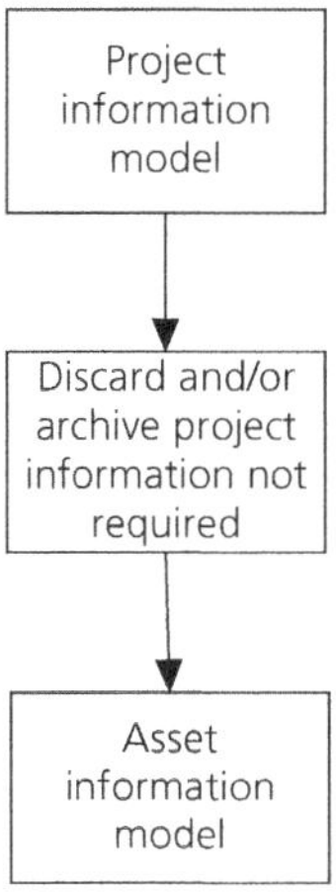

facility management systems and/or a spreadsheet data format that contains digital information about a built asset (e.g. COBie (Construction Operations Building Information Exchange) or similar). As noted earlier in this chapter, requirements as to the format of the BIM deliverables should have been agreed beforehand in the exchange information requirements.

In practice, the asset information model comprises all files and data that the delivery team gives to the appointing party during the handover at the completion of the project, and the appointing party/end-user uses that information to set up a facility management system to aid with the operation of the built asset in use; this is all part of the 'golden thread' of information (dealt with in Chapter 5 of this book).

3.6. Summary

- The purpose of the BIM strategy is for the appointing party to establish what it wants or needs to obtain in respect of information and/or data (known as the BIM deliverables) through the use of BIM.
- A statement of need is normally produced first in an attempt to define and describe the outline BIM requirements and deliverables of the appointing party.
- An independent BIM adviser appointed in the early stages of a project can help an appointing party to assess its long-term business objectives and to define how BIM should be used by the appointing party.
- Whereas the general project specification defines the nature of the built asset, the project BIM brief is a document that defines the appointing party's requirements and expectations for a project specifically with regard to BIM.
- The appointing party needs to appoint one or more lead appointed parties to operate the BIM procedures.

- The exchange information requirements define the information about the built asset that the appointing party wishes to procure and is normally divided into three parts, namely technical, management and commercial.
- To encourage standardisation and collaboration, it is important that standard BIM specifications/standards are established and used.
- The BS EN ISO 19650 (BSI, 2019) standard is an international set of standards of good practice for managing information over the whole lifecycle of a built asset using BIM.
- At an early stage of the BIM process, it needs to be determined (initially and principally by the appointing party) what BIM models are required.
- The project information model is a model that is developed during the design, production and construction phase (the delivery phase) of a project and consists of documentation, graphical and non-graphical information defining the delivered built asset.
- The asset information model is developed from the project information model and contains information related to the use, management, maintenance and operation of the built asset.

Peter Barnes
ISBN 978-1-83608-515-7
https://doi.org/10.1108/978-1-83608-512-620251005
Published under exclusive licence by Emerald Publishing Limited

Chapter 4
BIM parties and teams and the information management process

4.1. BIM parties and teams

The adoption of the BIM process requires a change to most if not all relationships on a project. In the case where contractual relationships are not appropriately defined, BIM-based collaborative work may give rise to legal liabilities; hence, a clear definition of the contractual roles and obligations of key participants will assist in regulating the defined responsibilities set out in the BIM execution plan (dealt with in Chapter 5 of this book).

The BS EN ISO 19650 (BSI, 2019) series of BIM standards refers to the appointing party, the lead appointed party, the appointed party, the project team, the delivery team and the task team; all these roles and teams, as well as the BIM coordinator, are commented on below.

It must be remembered that, at the present time, there are separate appointments between the appointing party and a lead appointed party, and between a lead appointed party and an appointed party. A diagram showing the party appointment relationship is shown in Figure 2.2 of Chapter 2.

However, at some time in the future, this typical bi-party chain of contracts may be (and hopefully will be) replaced by some form of collaborative multi-party form of contract or appointment.

Appointing party

The appointing party is the entity responsible for initiating the project (this party is often referred to as the client or the end-user). The appointing party (or a party acting on behalf of the appointing party) is a member of the project team. There is normally only one appointing party per project; the appointing party will appoint either one, or more than one, lead appointed parties (as shown in Figure 4.1).

The activities and required outputs of the appointing party can be summarised as being to

- establish the project's information requirements, information delivery milestones (data drops) and information standards
- identify specific procedures for the production of information including its generation, delivery and secure management

Figure 4.1 Lead appointed parties appointed by the appointing party

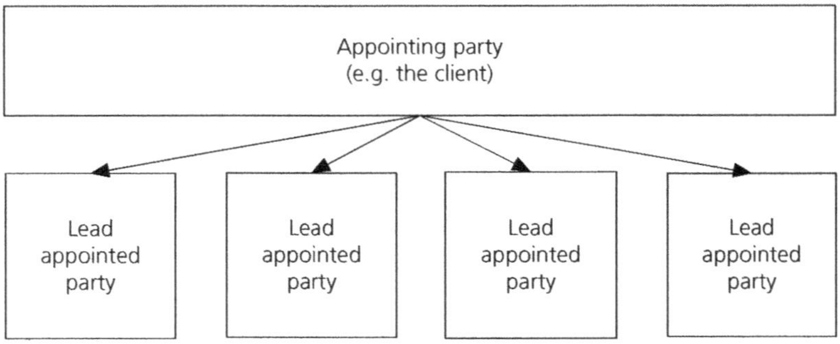

- identify existing information and/or resources that are relevant to the delivery teams who will subsequently be appointed to the project
- establish the project's information protocol (amended as necessary) for incorporation into all project appointments throughout the supply chain
- establish a common data environment to support the project and the collaborative production of information. The common data environment may be hosted, managed and supported by the appointing party or by a third party or parties.

The appointing party's outputs in compiling the tender packages for the lead appointed parties include

- the production of the exchange information requirements
- the provision of existing information and resources that are relevant to the tendering process
- the provision of details of how the tender will be evaluated
- the provision of the overall project requirements for information delivery, standards and processes
- the provision of the project's information protocol
- the creation of the project's information requirements and the identification of the project standards.

Throughout the project, the appointing party will review each delivery team's information model against the project information requirements and will accept or reject it as appropriate.

At the end of each stage of the process, and at the conclusion of the entire process, the appointing party will need to capture the lessons learnt on review with the lead appointed party (and any other relevant parties).

Lead appointed party

A lead appointed party is the party responsible for coordinating information between the delivery team (detailed below) and the appointing party. The lead appointed party is a member of the project team (detailed below) and of a delivery team. There can be and often is more than one lead appointed party, as there is often more than one delivery team (e.g. a designer may lead a design delivery team and a contractor may lead a contracting delivery team). The above possibility is set out in Figure 2.8 of Chapter 2.

A lead appointed party is appointed directly by the appointing party, taking the role of a coordinator between a delivery team and the appointing party, and can assign tasks to its own appointed parties. A lead appointed party will (normally) appoint several appointed parties to form part of the delivery team and/or task teams (detailed below) as set out in Figure 2.1 of Chapter 2.

The BS EN ISO 19650 (BSI, 2019) series refers to the lead appointed party in two ways

- the prospective lead appointed party
- the appointed lead appointed party.

The prospective lead appointed party

A prospective lead appointed party is a party that has not yet been appointed by the appointing party as a lead appointed party.

The key activities and outputs of a prospective lead appointed party are, in response to an invitation to tender and in collaboration with the relevant prospective members of the delivery team, to

- establish the (pre-appointment) BIM execution plan (dealt with in Chapter 5 of this book)
- summarise the delivery team's capability and capacity to manage and produce information
- establish the delivery team's mobilisation plan
- establish a high-level responsibility matrix to identify what information is to be produced, when and by whom
- create a risk register to deal with risks associated with the non-timely delivery of information, as may be identified in the tender documents
- compile the master information delivery plan (as dealt with later in this chapter) through the collation of the various task information delivery plans (as dealt with later in this chapter) generated by each prospective task team.

Appointed lead appointed party

An appointed lead appointed party is a party that has been appointed by the appointing party as a lead appointed party.

The (appointed) lead appointed party needs to compile appointment documents for each future appointed party. These should be tailored so that they include the exchange information requirements (for the task team) and the task information delivery plans that are relevant to the proposed appointments, alongside the delivery team's BIM execution plan and the agreed project level documents required.

The key activities and outputs of an appointed lead appointed party are, in collaboration with the relevant prospective members of the delivery team, to

- appoint parties to undertake the appointed party tasks
- mobilise all necessary resources
- ensure that the required training for the prospective appointed parties is in place
- ensure that the required hardware and software for the operation of the BIM process is in place
- establish the (post-appointment) BIM execution plan (dealt with in Chapter 5 of this book)
- refine and develop the high-level responsibility matrix to identify what information is to be produced, when and by whom (both in respect of the task teams and the appointed parties)
- update the risk register to deal with risks associated with the non-timely delivery of information as may be identified in the tender documents
- update the master information delivery plan through the collation of the various updated task information delivery plans generated by each prospective task team
- manage the progression of the delivery team's information model and, at the end of each information milestone (data drop), review and authorise each task team's information model to ensure that it meets the project's information standard, the BIM execution plan and the exchange information requirements as assigned to the applicable task team. A task information model which is found to be non-compliant (in any respect) should be rejected and the relevant task team(s) should be advised to amend their information accordingly
- capture lessons learnt, in collaboration with the appointing party, ideally throughout the term of the appointment, and also at the end of the appointment.

Appointed party

An appointed party is a party tendering for or actually appointed to a project. An appointed party is a member of the project team, the delivery team and one or more task teams.

An appointed party is the receiver of information concerning works, goods or services from a lead appointed party; most of the requirements below the level of the lead appointed parties are directed at a task team which comprises of various appointed parties (as set out in Figure 4.2).

The appointed party is referred to in two ways

- a prospective appointed party
- an appointed appointed party.

Prospective appointed party (i.e. an appointed party prior to appointment)

The key activities of a prospective appointed party (forming part of a prospective task team) are to

Figure 4.2 The task team's structure

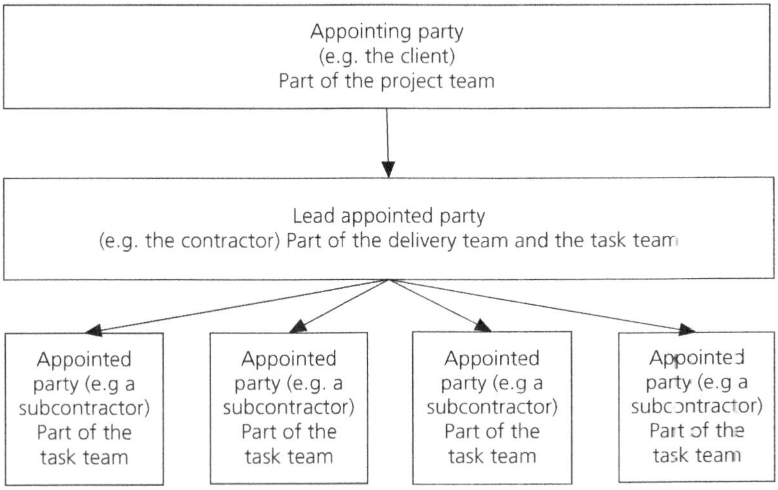

- assess its own capability and capacity in respect of producing and managing information, and ascertain the availability of its own appropriate and suitable IT (information technology) resources
- assist the lead appointed party in producing and confirming the delivery team's BIM execution plan
- notify the lead appointed party of any required additions or amendments to the project's information standard, production methods and procedures and to the information protocol. Any such additions or amendments will, of course, need to be agreed with the lead appointed party (and, in some cases, with the appointing party)
- assist the lead appointed party in refining, developing and detailing the high-level responsibility matrix in the BIM execution plan to ensure that it reflects what information is actually to be produced, at what time and by whom (i.e which task team and/or individual appointed parties)
- establish, in collaboration with the lead appointed party and the other prospective appointed parties across the task teams, the task information delivery plan(s) so that the master information delivery plan can be compiled.

(Appointed) appointed party (i.e. an appointed party after appointment)

The (appointed) appointed party (by now forming part of a task team) works closely with the lead appointed party to mobilise the task team's resources (personnel, hardware, software and IT) and to participate in training and education where this is found to be necessary to fill any knowledge and skills gaps.

The (appointed) appointed party collaboratively generates its information in compliance with the information standard, information production methods and procedures, using the appointing party's reference information and shared resources. To be able to generate appropriate information, the appointed party will need to understand the project definitions around the level of information need, the common data environment container breakdown structure, and an outline of what information is being produced by other appointed parties/task teams which may impact on its own activities.

The (appointed) appointed party checks the information containers that it produces to make sure that they are in accordance with the project information production methods and procedures and are in accordance with the project information standard. Any non-compliance discovered needs to be addressed by the party who originated the information. Once the procedural aspects of the information containers have been checked, the appointed party checks the contents of the information containers to make sure that the contents meet the information requirements and are in accordance with the level of information need.

The (appointed) appointed party then takes part with the relevant lead appointed party in a delivery team review of the project information model. This process may be repeated many times during the production of information, leading up to delivery of the project information model to the appointing party. At each point of this iterative process, the appointed party needs to make the agreed and necessary changes to its own information containers.

When the appointed party/task team's information model is ready to be delivered, it is submitted to the applicable lead appointed party for its review and authorisation. If the appointed party's information is rejected during this process, then this will be returned to the appointed party by the lead appointed party for further amendment and resubmission.

BIM coordinator

The BIM coordinator has a design role with particular responsibility for model coordination and clash detection.

Normally, on non-BIM projects, coordination and clash detection are part of the design activities undertaken by the lead designer; therefore, great care needs to be taken when appointing a BIM coordinator to undertake a similar role in respect of BIM projects.

The roles and responsibilities of the project BIM coordinator are to

- coordinate the various models produced
- coordinate the process leading up to the production of the models to ensure that the various models are compatible with one another
- establish the kind of models that need to be produced
- check what the purpose of the models is and consider how they will be created and how they will link in with each other
- maintain a collection of models from the various different sources
- make sure that the information on a model and between models is interoperable (i.e. so that data can be exchanged) and is up to date

- ensure that each appointed party has published a version of the required model for each significant milestone stage (data drop), as identified in the BIM coordination programme
- record and monitor shared data and relationships between the various models (e.g. grids, floor levels, shared project coordinates etc.)
- identify and agree any co-located or shared technical infrastructure needs, software package interoperability requirements and standards to be used by each delivery team member to deliver the BIM project
- create and administer the agreed document and model sharing/publication system
- coordinate the handover of the model and data at the agreed milestones (data drops) in line with the BIM coordination programme
- identify and document clashes between different appointed parties/delivery teams' models using clash detection software
- establish quality control procedures to check that all the models are accurate and that the level of detail in each model is as required and is fit for its intended purpose.

The project team

The project team comprises all parties (i.e. the appointing party, the lead appointed parties and the appointed parties) involved in the delivery process of a built asset using BIM.

The delivery team

The delivery team comprises all participants responsible for the creation of particular aspects of the required information. The delivery team comprises lead appointed parties and the relevant appointed parties. Depending on the contract type, there may be only one delivery team or (more usually) several delivery teams.

Task team

The task team comprises only those appointed parties (within a delivery team) tasked with performing a specific task going towards the delivery of BIM deliverables. There are normally many task teams.

The teams

The structure of the teams is as set out in Figure 4.2.

4.2. Activities and stages

To maintain structure and to maintain a standardisation of approach, various BIM activities forming stages of the BIM process are set out in BS EN ISO 19650-2 (BSI, 2018b); these are identified below as eight steps.

- Step 1 deals with assessment and need.
- Step 2 deals with the invitation to tender.
- Step 3 deals with the tender response.
- Step 4 deals with the appointment.
- Step 5 deals with mobilisation.
- Step 6 deals with the collaborative production of information.
- Step 7 deals with the information model delivery.
- Step 8 deals with the project close-out.

Steps 1 to 8 combined relate to a project as a whole and only occur once between an appointing party and a lead appointed party, whereas Steps 2 to 7 combined are repeated in respect of each lead appointed party's appointment of each appointed party.

Step 1: assessment and need

Step 1 is covered by clause 5.1 of BS EN ISO 19650-2 (BSI, 2018b). In this step, the appointing party determines why the required works are to be carried out (e.g. to improve a highway's capacity and safety) and then establishes the project information requirements, the information standards and the project milestones (data drops) necessary for the assessed need.

Step 2: invitation to tender

Step 2 is covered by clause 5.2 of BS EN ISO 19650-2 (BSI, 2018b). In this step, the appointing party establishes its tender response evaluation criteria, and then issues tender documents inviting prospective lead appointed parties to formally provide a tender for the works.

At a later stage, the lead appointed party follows a similar approach in inviting tenders for the works from prospective appointed parties.

Step 3: tender response

Step 3 is covered by clause 5.3 of BS EN ISO 19650-2 (BSI, 2018b). In this step, the prospective lead appointed parties prepare and submit to the appointing party a tender response demonstrating their approach to the works, setting out their capability and capacity through the applicable prospective delivery teams.

Similarly, and at a later stage, the prospective appointed parties prepare and submit to the relevant lead appointed party their tender responses demonstrating their approach to the works, setting out their capability and capacity in respect of the established requirements.

Step 4: appointment

Step 4 is covered by clause 5.4 of BS EN ISO 19650-2 (BSI, 2018b). At this stage, the various tenders from the prospective lead appointed parties are evaluated and the successful lead appointed party/parties are appointed by the appointing party. The required delivery teams are decided on, headed up by the applicable lead appointed party.

At a later stage, the various tenders from the prospective appointed parties are evaluated and the successful appointed parties are appointed by the relevant lead appointed party. The delivery team is put together by the applicable lead appointed party and then carries out detailed planning and a schedule outlining how works will be completed to meet the information requirements and milestones (data drops).

Step 5: mobilisation

Step 5 is covered by clause 5.5 of BS EN ISO 19650-2 (BSI, 2018b). After all the ground works have been laid in line with the steps above, the applicable delivery team ensures that it has the people, processes and technology available to deliver the works, and, if necessary, the delivery teams are provided with further training. As mobilisation is now to take place at this stage, the developed procedures and technology are tested and documented.

Step 6: collaborative production of information

Step 6 is covered by clause 5.6 of BS EN ISO 19650-2 (BSI, 2018b). At this time, and going forward, information and data are generated, coordinated, checked, reviewed and approved by the delivery team (by way of the lead appointed party) for publishing to the appointing party. The base information and data are generated by task teams and then, on approval, the information is shared to other task teams for their reference (e.g. a road design is shared with the lighting and signals task team) and is then collected together by the delivery team.

The output at this stage results in the creation of a project information model which is a collection of information and data that responds to the specific requirements set out in the appointing party's exchange information requirements (EIR) (and/or the lead appointed party's EIR, as appropriate) and which provides all the information required to carry out the delivery phase of the built asset. Examples of the outputs at this stage include computerised models, databases, spreadsheets and other documentation.

Step 7: information model delivery

Step 7 is covered by clause 5.7 of BS EN ISO 19650-2 (BSI, 2018b). At this stage the appointing party (and/or the lead appointed party, as appropriate) reviews the published project information model against the acceptance criteria and either approves or rejects it as a BIM deliverable. This exercise is subsequently repeated for each milestone (data drop) deliverable on the project.

Step 8: project close-out

Step 8 is covered by clause 5.8 of BS EN ISO 19650-2 (BSI, 2018b). Once activities and works have been approved, and the work on site is completed and the various appointment(s) come to an end, the information is archived and aggregated from the last produced project information model into an asset information model for the ongoing operation and maintenance of the built asset. In reality, the ultimate output of the information management process is the creation of an asset information model which provides all the information required to enable the appointing party (or end-user) to perform the operational phase of the built asset. At the end of the project, some components of the project information model may be transferred to the asset information model, while unnecessary and surplus data is archived. The asset information model can include graphical models, non-graphical data and all necessary documentation for the ongoing maintenance, operation and management of the built asset, and forms part of the continuing 'golden thread' (dealt with in Chapter 5 of this book).

Summary of the activity steps

Following the initial assessment and need exercise carried out by (or on behalf of) the appointing party, the information management process is characterised by three phases, as outlined below

- the procurement phase – comprising Steps 2 and 3 above
- the planning phase – comprising Steps 4 and 5 above
- the production phase – comprising Steps 6 and 7 above.

The above is shown diagrammatically in Figure 4.3.

Figure 4.3 The BIM activities steps

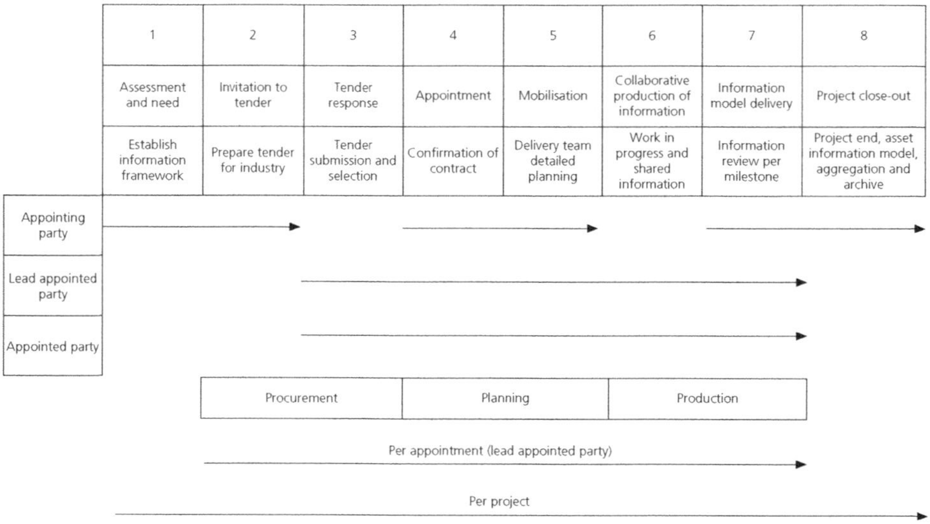

4.3. Assessment and need

As noted above, at the initial stages of the BIM process, there is a requirement to establish the appointing party's BIM requirements, allocate the respective information management functions, consider the benefits of and then establish a common data environment, review the digital security requirements that are needed and ensure that open and interoperable BIM solutions are available.

The appointing party determines what works are to be carried out, particularly in respect of the BIM process, and also establishes the project's information framework; this will normally consist of the following

- the information requirements (the exchange information requirements, the organisational information requirements, the project information requirements and the asset information requirements)
- the required project information standards
- the required information delivery milestones (data drops)
- the required project information production methods and procedures
- the required reference information and shared resources elements
- the project's common data environment (CDE)
- the project information protocol.

However, to be fully effective, before any of the above can be finalised, the appointing party must first appoint the project's BIM information manager.

The BIM information manager is one or more individuals from within the appointing party's own organisation or from a third party who will lead the information and data management function for the project.

BIM information manager

The BIM information manager is essentially a procedural gatekeeper, policing the model(s), the common data environment (CDE), the information and data, and the associated processes, to ensure that they follow the requirements of the various appointments and also to ensure that the data that is held is accessible but secure.

The BIM information manager appointment may change through the course of the project (e.g. a lead designer or a lead consultant may be the BIM information manager during the early stages of the project, with a contractor then becoming the BIM information manager during the construction phase) or may remain the same person/organisation throughout the project period. There is a school of thought that the latter situation is preferable as this ensures consistency and continuity in approach.

It must be clearly understood that the BIM information manager is not a BIM coordinator and, consequently, has no responsibility for clash detection or model coordination; the BIM information manager does not need to be a designer (although sometimes is).

The BIM information manager is purely responsible for the management of information, information processes, compliance with agreed procedures and coordination of data and so on, not with the coordination of the design itself.

While the precise scope of the BIM information manager's function is to be determined by the appointing party, the role of the BIM information manager would normally be to

- establish a common data environment (CDE) and then host, maintain and validate the information flow during the project
- establish the basis of a BIM execution plan – that is, how the BIM requirements will be executed in practice – and prepare and assign resources (people, processes and technology) within the delivery team. Sometimes (but by no means always) the BIM information manager is also responsible for creating contract appointment documents for each lead appointed party and other appointed parties
- initiate and implement a master information delivery plan, a project information plan and an asset information delivery plan
- be responsible for data security and also for creating a data archive
- manage and supervise the progress of the project information model being created by the delivery teams
- check whether the project information model that is being provided meets the requirements as set by the appointing party
- capture lessons learned with the appointing party and each of the lead appointed parties.

Assuming that there are multiple models, which will normally be the case, the BIM information manager would normally be required to

- determine who will assemble the composite federated model
- establish exchange methods and identifications for files and formats
- locate a central, shared digital repository for the drawings and other relevant documents
- establish the origin point for the geographic model alignment
- implement regularly scheduled reviews and quality checks of the various models that are being produced
- ensure that the constituent parts of the project BIM model(s) are compliant with the master information delivery plan, the project information plan and the asset information plan (as appropriate).

Although there may be several subsidiary models in existence, it is during the process of coordinating the data, which is a process that is led by the BIM information manager, that the models are usually linked (or referenced) together into one federated BIM model.

Normally, during the design phase the model or models will be updated as each consultant organisation or other BIM participant (i.e. lead appointed party and/or appointed party) issues its next release of design information. However, during construction, the entire project team, including the designers, the contractor and the major subcontractors, need to be involved and need to be aware that the models are being maintained and updated. Therefore, the incorporation of design changes into the BIM model(s) must be an identified and an identifiable process controlled by the BIM information manager.

After the BIM information manager has been selected, the appointing party can define the project's information requirements framework.

Information requirements

The efficient and effective collaborative production of information starts with clear and well-defined information requirements.

It is difficult and inefficient to plan what information is to be created and delivered (and the time and resources needed to do so) without first understanding what information is needed.

BS EN ISO 19650 (BSI, 2019) places an onus on the appointing party to understand and specify what project and asset information is required in respect of the BIM process at the start of the project, and then to communicate that information to the relevant lead appointed party.

Effective information requirements should include the following

- who the information is for
- why the information is needed
- the form (i.e. how it is presented) and the format (i.e. how it is encoded) of the required information
- the level of information need – that is, the required quantity and granularity of information (bearing in mind that only the information that is actually needed should be produced, as surplus material is a waste)

- the acceptance criteria – that is, how the information provided is to be reviewed for acceptance
- the supporting information, documentation, standards or resources that are required to assist with information production
- the exchange date (the data drops) – that is, when the required information is to be delivered.

BS EN ISO 19650 defines four types of information requirements

- organisational information requirements (OIR): organisational objectives, needs and necessities
- project information requirements (PIR): information requirements for the delivery phase of an asset
- asset information requirements (AIR): information required for the ongoing operation and maintenance of an asset
- exchange information requirements (EIR): how, when and what information will be delivered. The exchange information requirements are derived from the organisation information requirements, the project information requirements and the asset information requirements.

The high-level information requirements (which apply to both the organisational information requirements and the project information requirements) establish the overall purpose of the works and define why information and data are needed.

The detailed information requirements (which apply to both the exchange information requirements and the asset information requirements) specify how information should be produced and exchanged.

The above structure of information requirements is encapsulated in Figure 4.4.

Organisational information requirements (OIR)

The organisation information requirements are determined by the appointing party and contain the high-level requirements that define the information needed to meet the particular appointing party's strategic business objectives and the needs of its own asset management system. Organisational information requirements ensure the correct information feeds back into an organisation to support wider business decisions (for example, the information needed by a rail operator is very different from that of a manufacturing facility). The organisational information requirements are the starting point from which of all the other information requirements are defined.

Project information requirements (PIR)

The project information requirements are again determined by the appointing party and these contain the high-level project-specific information requirements, production methods and procedures required at specified key decision points for the delivery phase of a built asset.

Figure 4.4 Structure of the information requirements

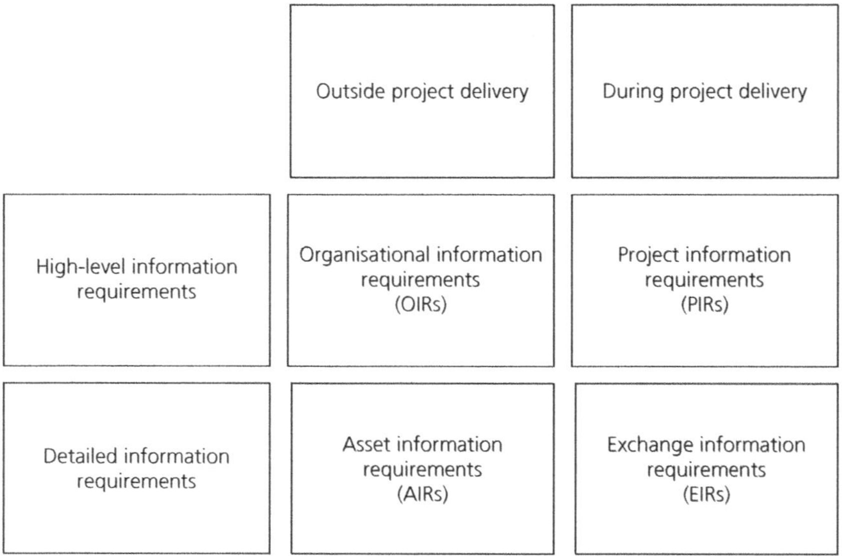

The following points need to be taken into consideration and/or need to be stated when establishing the project information requirements

- the project scope, giving the basic information about the built asset/project
- the intended purpose for which the information obtained will be used by the appointing party
- the project plan of work – that is, how (or if) the project will be broken down into stages or intervals
- the intended procurement route, including how appointments will be structured, the relationships between the various parties and the rules that govern the project
- what standardised elements for exchanging information will be established in the exchange information requirements
- what structuring and classification systems(s) of information (e.g. Uniclass 2015 or similar) will apply
- the method to be used for specifying the level of information need
- how information during the operational phase will be used and what standardised elements for operational use have been established for the project
- how existing information will be captured
- how new information will be generated, reviewed or approved
- details of the measures relating to the security or distribution of information
- how the information required is to be delivered to the appointing party.

An example of the project information requirements is set out below.

1. Project scope

   Provide a description of the project, specifying objectives, purpose and any relevant timeframes.

   1.1 Project purpose

   Provide an explanation of why the project is required by the organisation/appointing party, including the intended use and occupation, if applicable.

   1.2 Project objectives

   Outline the project goals and objectives, focusing on how the project can contribute to achieving the organisational information requirements (OIR) strategic objectives.

   1.3 Project details

   Provide the project details, which should include: site address, procurement method, internal project team and the expected date for the facility to be in operation.

   1.4 Legacy information

   Input any reference to any existing information related to the proposed built asset that better informs the project information requirements. Also, include hyperlinks or reference names/number and confirm the location of the legacy information.

2. Plan of works

   This section identifies the plan of works for the project. The plan of works generally describes the project/work stages that are used to divide the project, which may also be used for contractual purposes.

   2.1 Information delivery milestones (data drops)

   Define the information delivery milestones (data drops), where and when specified information deliverables are exchanged during the project. An information delivery milestones plan and requirements schedule/table should be provided.

3. Information requirements

   This section is used to define the information requirements that are specific to the project and describe how they are managed. These should be the specific requirements of the delivery phase of an asset.

   3.1 Project information requirements

   Outline the project information requirements, which should be derived from the organisation's strategic objectives.

Asset information requirements (AIR)

The asset information requirements are information requirements that relate to the operational phase of a built asset. These are detailed information requirements that outline the information needed to operate and maintain the built asset throughout its lifecycle. The AIRs define what information is required and how it should be delivered for the asset information model, keeping in mind that one of the main aims of the asset information model is to ensure that the right people have access to the right information at the right time to effectively maintain and operate the built asset.

Exchange information requirements (EIR)

The EIRs are detailed appointment-specific requirements that precisely specify what information is needed at each information milestone (data drop) to enable parties to complete necessary activities during the delivery and operation phases of a built asset. EIRs are collectively defined from the organisational information requirements, the project information requirements and the asset information requirements, and are developed for each appointment. The appointing party sets the EIRs for their appointment with the lead appointed party, while the lead appointed party sets their own (specific and possibly different, but consistent with the original EIRs) EIRs for each appointment with their appointed parties.

An example of an EIR could be a spreadsheet detailing the types of information that need to be delivered at each project milestone (data drops), including the required forms and formats for information.

The relationship between the various types of information requirements is encapsulated in Figure 4.5.

Level of information need

The level of information need is defined by the appointing party for each information deliverable, as outlined in both the project's information standard and the exchange information requirements.

Figure 4.5 Relationship between different types of information requirements

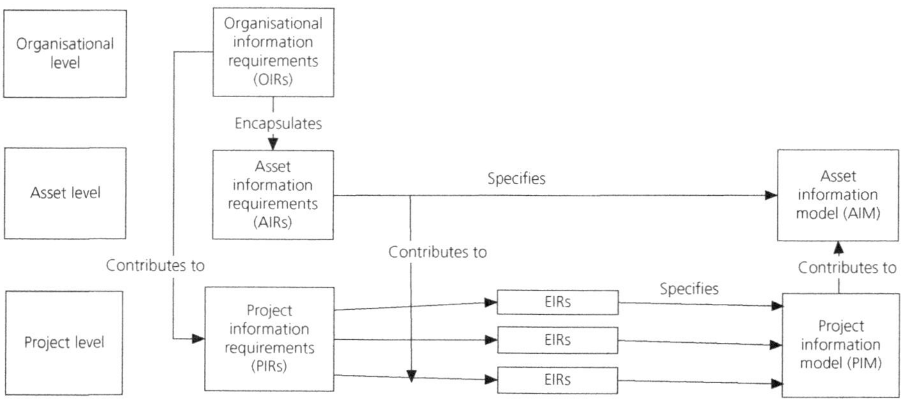

A key principle of BS EN ISO 19650 (BSI, 2019) is the minimisation of 'information waste'. Producing too much information is a waste of time and resources, while producing too little can lead to ill-informed decisions and mistakes.

To avoid information waste, the appointing party must clearly state their level of information need, namely the minimum (and in some cases, maximum) quantity and granularity of information required to meet its needs.

In this context, information requested and provided should not

- exceed the required level of information need
- extend beyond the scope
- lack a specific purpose for its use
- duplicate information created by other task teams
- contain superfluous details.

Information standard

The project's information standard is a document formed by the appointing party that specifies how and by what standard all information is to be produced. The goal is to ensure a consistent approach for the structuring and classification of information that is most appropriate for the project. The information standards document may include details of languages, units, origins and orientations, coordinate systems, naming conventions, codifications, and text, dimension and line styles and so on.

Information delivery milestones (data drops)

A project's information delivery milestones (data drops) are set by the appointing party and these milestones specify when information is to be exchanged on a project. The dates of milestones should be developed from the project information requirements and should reflect the required project key decision points and the information deliverable(s) required for each of them.

Information production methods and procedures

The project's information production methods and procedures is a document (or a set of documents) produced by the appointing party that details the processes for how work should be carried out by the delivery team. This may include how information should be

- generated, reviewed and approved
- secured and shared
- exchanged by and with the appointing party.

Reference information and shared resources

The appointing party should share supporting information to assist in the production and management of information. Examples of reference information and shared resources can include

- existing asset information
- references to relevant industry standards
- supporting documents or guidance material.

Establish the project's common data environment (CDE)

The appointing party must also establish the project's common data environment (CDE). The CDE is a central electronic repository/store used to collect, manage and disseminate information (and is dealt with in more detail in Chapter 5 of this book).

4.4. Summary

- The BS EN ISO 19650 (BSI, 2019) series of BIM standards refers to the appointing party, the lead appointed party, the appointed party, the project team, the delivery team and the task team.
- The appointing party is the entity responsible for initiating the project (sometimes referred to as the client or the end-user).
- A lead appointed party is appointed directly by the appointing party; it takes the role of a coordinator between a delivery team and the appointing party and can assign tasks to its own appointed parties.
- An appointed party is a party tendering for or actually appointed to a project.
- An appointed party is a member of the project team, the delivery team and one or more task teams.
- The BIM coordinator essentially has a design role with responsibility for model coordination and clash detection.
- There are various teams in the BIM process, the most important of which are the project team, the delivery team and the task team.
- The BIM process comprises eight steps (as extracted from BS EN ISO 19650-2 (BSI, 2018b)), and is normally broken down into a procurement phase, a planning phase and a production phase.
- The BIM information manager is essentially a procedural gatekeeper, policing the model(s), the common data environment (CDE) and the associated processes.
- Key information requirement documents are: the organisational information requirements, the project information requirements, the asset information requirements and the exchange information requirements.

Peter Barnes
ISBN 978-1-83608-515-7
https://doi.org/10.1108/978-1-83608-512-620251006
Published under exclusive licence by Emerald Publishing Limited

Chapter 5
The procurement phase, the planning phase, the production phase and the project close-out phase

5.1. Invitation to tender

Once the appointing party has defined its information framework, it can then establish the tender response requirements and evaluation criteria, and issue tender documents inviting prospective lead appointed parties to formally tender for the works.

The tender response requirements are established by the appointing party and those requirements outline the minimum conditions that the prospective lead appointed party is to meet and demonstrate in its tender response.

The appointing party also establishes the degree to which it will evaluate tenderers in meeting these requirements. This is known as the tender evaluation criteria.

The goal of the tender response requirements and the tender evaluation criteria is to ensure a consistent and fair approach to tender selection.

Once finalised, these requirements and evaluation criteria are compiled, and the tender invitation documents are issued to the prospective lead appointed parties. The tender invitation documents also include the project information protocol (previously known as the BIM protocol) (which is a document which explains how the BIM process is to be incorporated into appointments/contracts), together with its compiled requirements and standards.

After this, in the next phase, an invitation to tender is issued to those that will form part of the overall BIM delivery teams (i.e. the prospective appointed parties).

The invitation to tender to a lead appointed party or to an appointed party needs to include

- the project BIM brief
- the exchange information requirements
- the organisation information requirements
- the project information requirements
- the asset information requirements.

If the appointing party has its own BIM execution plan template that complies with BS EN ISO 19650-2 (BSI, 2018b), that template should be made available as part of the invitation to tender documentation. If not, a BIM execution plan will need to be created.

5.2. Tender response

The prospective delivery teams prepare a tender response demonstrating their approach to the works and the BIM process as well as their capability and capacity for achieving the BIM requirements. The tender response provides a basis for the appointing party to evaluate and then select the lead appointed party who is most suitable to carry out the works and comply with the BIM requirements and, in turn, a basis for the relevant lead appointed parties to evaluate and select the applicable appointed parties.

The tender response is a document submitted by prospective lead appointed parties (and prospective appointed parties) and contains the following

- a pre-appointment BIM execution plan
- a risk register.

Pre-appointment BIM execution plan (BEP)

The pre-appointment BIM execution plan (BEP) is a document produced by prospective lead appointed parties and their anticipated appointed parties. The document generally includes the standards, roles and responsibilities, and protocols that are used as a basis for a BIM project. It demonstrates the overall approach for model generation, information production, collaboration and exchange, and BIM deliverables, and outlines the prospective delivery team's proposed approach, capability and capacity to meet the appointing party's exchange information requirements.

The appointing party, together with its advisers, should check that the prospective appointed party's pre-appointment BIM execution plan satisfies the requirements of the exchange information requirements.

Establishing the pre-appointment BIM execution plan should involve collaboration between a lead appointed party and prospective appointed parties (where known) so that it reflects what the delivery team as a whole will do, not simply what the lead appointed party will do.

The pre-appointment BIM execution plan (BEP) should include

- a proposed information delivery strategy
- a proposed federation strategy
- a high-level responsibility matrix
- a proposed mobilisation plan
- a capability and capacity assessment. The pre-appointment BIM execution plan must demonstrate that each task team member has both the capability and the capacity to produce and manage information and data in line with the exchange information requirements
- a project implementation plan setting out the capability, competence and experience of potential appointed parties bidding for a project, along with the associated and necessary quality documentation

- the goals for collaboration and information modelling
- the project milestones (data drops) in line with the project programme (i.e. times when collected data will be accumulated and reported on)
- the deliverable strategy for the planned use of the information and data at every stage of the project (this strategy is often further developed during the course of the project works).

The (pre-appointment) BIM execution plan also provides an opportunity for the prospective lead appointed party (in conjunction with the relevant prospective appointed parties) to identify additions and/or amendments to the project's production methods and procedures and its information standard. This may be needed so that

- information and data can be effectively generated, reviewed, approved, authorised and exchanged by the different parties involved
- the distribution and delivery of information is secure and safe
- an assessment can be made (by the lead appointed party, with input from the appointed parties) of a task team's capability and capacity (e.g. software access and use, space, object type and object component naming conventions, hardware, software and information technology infrastructure availability and capacity)
- the delivery team's capability and capacity may be established by the lead appointed party.

Refer to overview provided in Figure 5.1.

Delivery team's mobilisation plan

The delivery team's mobilisation plan needs to be established (by the applicable lead appointed party with input from the applicable prospective appointed parties).

The mobilisation plan serves two purposes.

- It informs the appointing party of the lead appointed party's approach to information management mobilisation.
- It is a tool for the lead appointed party to sufficiently plan out its mobilisation phase for information management after appointment.

The mobilisation plan may include

- export file formats, which obviously need to be consistent and coherent
- details testing that the delivery team's common data environment solutions perform as expected and that all appointed parties and task teams can access the common data environment appropriately
- details testing how separate information containers may manually or automatically be exchanged between solutions, depending on workflow, to reduce review waiting times
- ensuring site office information and communications technology infrastructure, including network lines, are in place to allow site staff to access information

Figure 5.1 Elements of the pre-appointment BIM execution plan

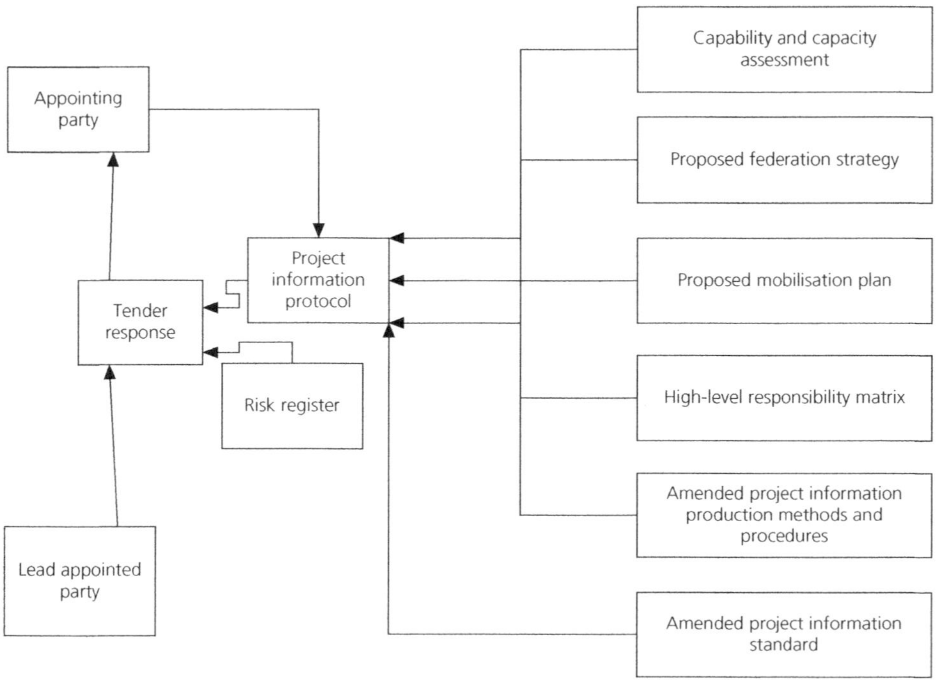

- production of information management plans or guides to help users operate the common data environment solutions
- the means of educating the delivery team on the project goals and the journey to get there; training the delivery team on the use of any mandatory solutions being provided by the appointing party
- the means of recruiting members who play a vital role in the management of information, including document/design managers and consultants
- support for individuals and organisations who join the delivery team during the appointment process.

Project implementation plan (PIP)

The project implementation plan details the delivery team's capability in summary form, incorporating the supplier BIM assessment form(s), the supplier information technology assessment form(s) and the supplier resource assessment form(s).

Risk register

At this stage, and after the completion of the pre-appointment BIM execution plan, the delivery team's risk register needs to be established by the applicable lead appointed party. An integrated risk register includes information management risks alongside other appointment-related risks;

both of these should be combined into a single risk register (and, ideally, to avoid confusion and confliction, there should not be separate supplementary risk registers). The risk register identifies all risks associated with the prospective delivery team's management and production of information throughout the appointment, as well as how they intend to manage those risks.

Risks to consider adding to the delivery team's risk register could include the following

- assumptions – for example, the existence, gaps and adequacy of the exchange information requirements received and the steps needed to address any concerns; the impact of roles and responsibilities for information management and how appointing party information requirements are being captured
- milestones – for example, consider whether there is sufficient capability and capacity and allowance for procurement lead times to meet the delivery milestones (data drops)
- information protocol – for example, consider whether the rights and responsibilities in the information protocol are acceptable from legal and commercial perspectives
- information delivery strategy – for example, consider whether an appropriate common data environment has been established and whether it has the capability to deliver the information delivery strategy
- methods, procedures and information standards – for example, consider the proposed methods and procedures against existing organisation and/or internal procedures
- mobilisation, capability and capacity – for example, ensure that consideration for lead times is included in mobilisation to allow for the common data environment procurement, internet connections and the provision of training providers.

5.3. The planning phase

During the planning phase, the delivery team carries out detailed planning, prepares a schedule outlining how works will be organised and managed, and mobilises resources (people, processes and technology) to meet this schedule.

5.4. Making appointments

Tenders are evaluated by the appointing party and the successful lead appointed party is selected. The delivery team (led by the applicable lead appointed party) then carries out detailed planning and prepares a schedule outlining how works will be completed.

Post-appointment BIM execution plan (BEP)

In response to the relevant exchange information requirements, lead appointed parties produce a post-appointment BIM execution plan.

A BIM execution plan is a document that outlines how BIM will be utilised on a specific project. It defines project objectives, the scope of BIM use, roles and responsibilities, information exchange protocols, software standards and project-specific requirements.

After appointments have been made, the (post-appointment) BIM execution plan forms the framework of the management and operational functions of the BIM process throughout the project's lifecycle. In other words, the BIM execution plan acts as a 'digital method statement' setting out the supply chain's agreed approach to BIM implementation on a project.

The post-appointment BIM execution plan sets out how the information required in the exchange information requirements will be provided and normally deals with

- management – roles, responsibilities and authorities
- project milestones (data drops) in line with the project programme
- deliverables strategy
- survey strategy
- existing legacy data use
- approval of information
- authorisation process
- revised project implementation plan (PIP) confirming the capability of the supply chain (the appointed parties)
- agreed processes for collaboration and modelling
- agreed detailed responsibility matrix
- master information delivery plan setting out when project information is to be prepared, by whom and using which protocols and procedures
- task information delivery plan setting out responsibility for delivery of each appointed party's information through the delivery team
- the file naming convention applicable
- the layer naming convention applicable
- any construction tolerances applicable
- drawing sheet templates
- annotation, dimensions, abbreviations and symbols details
- software exchange formats
- process and data management systems.

The (post-appointment) BIM execution plan should be developed with, and agreed by, each appointed party in a delivery team so that it is representative of the overall delivery team's proposals, not just that of the lead appointed party. The (post-appointment) BIM execution plan is a living document, and therefore should be continually updated as the project progresses.

Responsibility matrix

The responsibility matrix identifies who is responsible for each information management activity. It is used to define and allocate project tasks and responsibilities among team members. It plays an essential role in the BIM execution plan, outlining how a project will be executed and managed. The responsibility matrix identifies who is responsible for producing specific information or models and when they should do so. There are two main types of responsibility matrices

- high-level responsibility matrix: this matrix outlines the allocated responsibility for each main element of the information model and the associated key deliverables
- detailed responsibility matrix: this matrix is developed during the pre-appointment BIM execution plan and provides a more granular and detailed breakdown of responsibilities. This matrix specifies precisely what information is to be produced, when the information is to be exchanged and with whom, and which particular task team is responsible for its production, all in accordance with the specified information standard.

The lead appointed party refines the initial high-level responsibility matrix into a detailed responsibility matrix. This detailed responsibility matrix establishes a set of exchange information requirements for each appointed party, identifying

- what information is to be produced
- when the information is to be exchanged
- who the information is to be exchanged with
- the task team responsible for its production.

A RACI (responsible/accountable/consulted/informed) matrix (as shown in Figure 5.2) can be used for this; it outlines the task teams who are responsible, accountable, consulted and informed for each required information deliverable.

Task information delivery plan (TIDP)

A task information delivery plan is established by a lead appointed party and is a detailed plan for how a particular task team is going to deliver the information it has been asked to provide (models, documents, schedules, calculations, data and so on) in line with the detailed responsibility matrix. There is no prescribed format for the task information delivery plan, and it could be presented as a table, a spreadsheet, a database or something similar. The purpose of the task information delivery plan is to enable the task team to clarify what they will and will not deliver as part of their scope and what information will need to be exchanged between them and other task teams to allow timely coordination and progress across the delivery team. The purpose of the task information delivery plan is to identify specifically what information will be produced, which task team will produce it, how long it will take to produce and when it will be delivered.

Figure 5.2 The RACI matrix

	Task Team 1	Task Team 2	Task Team 3	Task Team 4	
Deliverable 1	**R**				Responsible
Deliverable 2		**A**			Accountable
Deliverable 3			**C**		Consulted
Deliverable 4				**I**	Informed

Master information delivery plan (MIDP)

The master information delivery plan (MIDP) is established by the lead appointed party; it lists information deliverables and sets out when project information is to be prepared, by whom, and utilising which protocols and procedures for each stage of the project.

The master information delivery plan is the primary plan for the preparation of project information which is set out as being required (the deliverables) in the exchange information requirements. Deliverables include (among other things) models, drawings, specifications, equipment, schedules and room data sheets, with some form of version change control being used across the document set. The master information delivery plan incorporates all the delivery team's individual task information delivery plans as may be identified in the information particulars and/or as provided in accordance with the information standard.

The purpose of the master information delivery plan is to allow the lead appointed party to check the delivery plans across the different task teams to make sure that they fit in with the overall delivery team schedule of activities. A delivery team's master information delivery plan obviously needs to be kept up to date with any changes that may be made to the individual task information delivery plans that form part of it; it also needs to be updated to include any task information delivery plans from newly engaged appointed parties or newly engaged task teams joining the delivery team.

Development of BIM execution plan

At this stage, the delivery team's BIM execution plan needs to be confirmed by the appointing party, the lead appointed party and the appointed parties. The BIM execution plan should be developed and agreed with each appointed party to ensure that it reflects their activities and their use of IT, and that by using the BIM execution plan they can work in accordance with the overall delivery team requirements.

Development of the content of the delivery team's BIM execution plan may also require engagement with the appointing party to discuss and agree any necessary additions or amendments to the project's information standard or the project's information production methods and procedures.

It is also important to make sure that the BIM execution plan meets the minimum content requirements of the appointing party. A BIM execution plan is likely to evolve over the life of the delivery team as additional parties are appointed. The lead appointed party is responsible for maintaining the delivery team's BIM execution plan so that it continues to represent the delivery team's information management approach. As the BIM execution plan will almost certainly be an appointment (contract) document, it will need to be subject to a process of formal change control with changes agreed with the appointing party and the appointed parties.

The lead appointed party's exchange information requirements need to be developed by the lead appointed party, based on the appointing party's own exchange information requirements. A set of exchange information requirements needs to be developed for each appointment, as the information requirements for each appointment is likely to differ from one to another (although it is normally based on the appointing party's exchange information requirements).

Therefore, each exchange information requirements document needs to detail the information required by the lead appointing party from the relevant appointed party.

The lead appointed party's appointment documents should be completed and executed by the appointing party. The lead appointed party's appointment documents should include the appointing party's exchange information requirements, the project's information standard, the project's information protocol, the delivery team's BIM execution plan and the master information delivery plan.

Following on from this, and in a similar vein and attaching similar documents (but in this case also the task information delivery plan), the various appointed parties' appointment documents need to be completed and executed by the lead appointed party.

The above is encapsulated in Figure 5.3.

5.5. Mobilisation stage

After appointments have been made it is now necessary for mobilisation to begin. The mobilisation is undertaken by both the lead appointed party and the individual appointed parties. The mobilisation stage provides the delivery team with an opportunity to ensure they have the people, processes and technology in place to deliver the information and to ensure that all

Figure 5.3 Elements of the post-appointment BIM execution plan

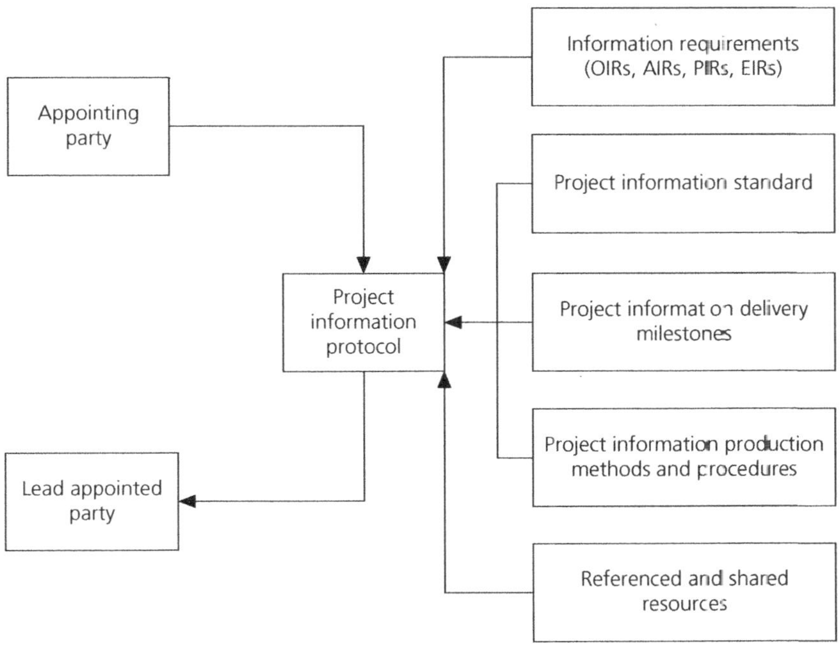

teams are suitably trained before work commences. Also, the information production methods and procedures, as well as information technologies (including the common data environment), need to be tested and documented.

The mobilisation stage involves

- confirmation of the resources availability of each task team and the engagement of additional appointed parties if additional capacity is needed
- development and delivery of educational resources (e.g. produce information management guides and share information standards)
- development and delivery of training, thereby ensuring all individuals can generate and/or manage information
- ensuring that all appointed parties and task teams can access the common data environment (CDE) suitably and without undue restriction.

5.6. Collaborative production of information (and the CDE)

In this stage information is generated, coordinated, checked, reviewed and approved by the delivery team for publishing to the appointing party. Information is generated by task teams and then, on approval, shared to other task teams for reference (e.g. a road design by the road design task team is shared with the lighting and signals task team).

The common data environment (CDE)

At the heart of the BIM process is a collaborative process for producing, managing and sharing information and digital models. This process requires a collaborative and accessible electronic platform on which this can be carried out. This is, in effect, what the common data environment (CDE) is. The CDE is the technology, including a data management system, that enables the BS EN ISO 19650 (BSI, 2019) information management process to be effectively dealt with.

A CDE is a system for collecting, filing, managing, sharing and distributing electronic data and other information. Most commonly a CDE is simply held electronically, and in such a case is simply a digital data room where geometric and other information from the contributing designers and other participants in the project come together. It is, in effect, an electronic filing system which ensures data consistency, version control and access control among project stakeholders, and from where the data and information can be extracted and shared with other members of the multi-disciplinary team (of consultants, contractors and subcontractors etc.) through a managed and controlled process.

The process of collecting, filing, managing, sharing and distributing data and other information was of course already and previously well established in a paper-based filing system. However, with the adoption of electronic technologies and with the massive increase of data produced on a typical construction and/or infrastructure project, the need for the good, efficient and effective collection, filing, management, sharing and distribution of electronic data and other information is clearly vitally important and is easier to achieve through an electronic filing system.

The CDE principles are well established and have been derived from mature project management methods amended to meet the specific needs of construction and infrastructure projects operating under the BIM process.

The CDE therefore provides a means of achieving a collaborative working environment between all participants in a construction or infrastructure project, particularly those operating under the BIM process, and helps to avoid duplication of information and mistakes in the exchange of data.

The principle behind the effective production and use of information and data is that information

- should be produced once and reused many times (to eliminate rework)
- should be continuously shared and trusted
- should have a clear auditable history, identifying where the information came from, who amended it and when
- should be checked, reviewed and approved before it is shared with other members of the delivery team
- should be clearly identifiable by its state, revision and status, and should have access controlled only as necessary and accordingly.

BS EN ISO 19650 (BSI, 2019) uses the term 'information container' to denote a part of the CDE where files, models, documents, datasets and so on are stored. Each information container should have a unique ID reference number in accordance with the agreed naming convention; BS EN ISO 19650 places the onus on the appointing party to implement and manage a CDE for the whole project. However, in practice, and for expediency, delivery teams may well operate their own CDEs in separate (but integrated) systems.

The CDE workflow organises the flow and management of information across the whole lifecycle of a built asset across four states: work in process, shared, published and archived. See chart in Figure 5.4.

A CDE could be stored on a project server, an extranet or a file-based retrieval system. However, increasingly, and given the electronic size of the data to be stored, CDEs use cloud-based software to provide the digital space to hold and share the information; being cloud based this can often make the sharing of data easier.

It is often argued that the appointing party should establish and host the CDE as they will often generate asset and/or project information before any consultants and/or contractors are appointed, and sometimes even before a BIM information manager has been appointed. Also, the information held in the CDE may initially have been obtained by the appointing party from the details of an existing asset (where one exists).

During the design and construction phases of an asset, the CDE forms the basis of the project information model and, when the asset is handed back over to the appointing party, the CDE

Figure 5.4 CDE workflow

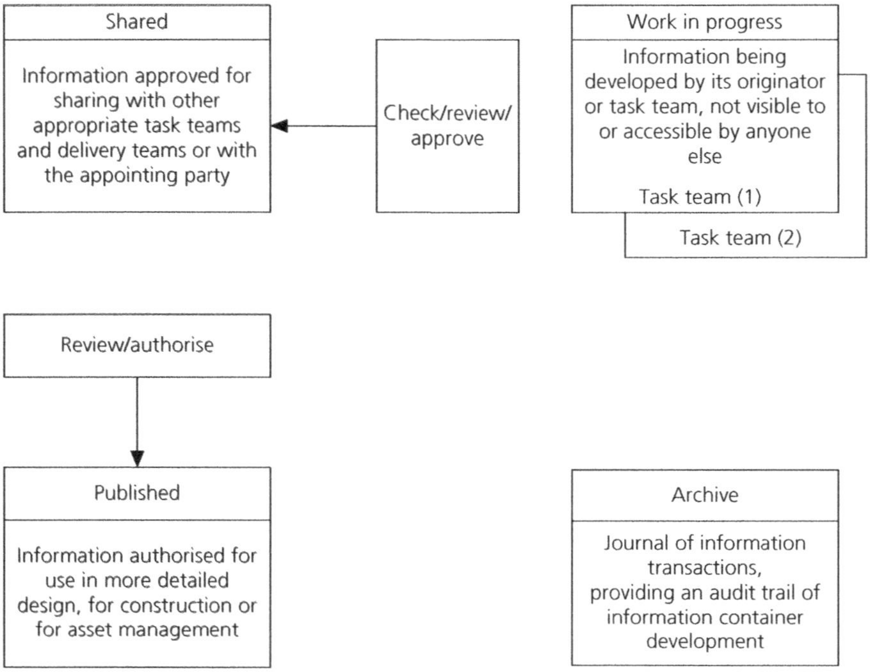

forms the basis of the asset information model. It is the asset information model that will ultimately be used by the appointing party for the operation of the built asset; therefore, if the appointing party establishes and hosts the CDE, this can also help avoid continuity problems that can occur if, rather than this, the hosting of the CDE is transferred from one organisation to another during the lifecycle of the built asset.

While it may not always be possible for an appointing party to store the CDE, not least because not every appointing party would have the capabilities (or the desire) to establish and host the CDE, if the appointing party does establish and host the CDE, then it is very important that the information provided and stored is maintained in a way that will be consistent with the way that information will be provided and stored at the later stages of the built asset's lifecycle.

Consequently, the file formats and file naming conventions established at the very beginning of the process must allow for the fact there will be a great deal of data stored in or extracted from the CDE throughout the lifecycle of the built asset, and that data therefore needs to be readily accessible and easily located through a data filing naming convention that has adequate capacity in terms of size and which is understood by all participants in the process.

Despite the advantages of the appointing party hosting the CDE, as outlined above, it is not (by any means) always the appointing party that establishes and hosts the CDE.

Often this matter is decided on primarily practical factors – sometimes simply being that the party that hosts the CDE has the technical capability in addition to sufficient knowledge and experience of BIM technology to ensure that the project is able to progress smoothly from this perspective.

The hosting of the CDE was an issue in the *Trant Engineering Limited v. Mott MacDonald Limited* (2017) court case. Trant was the design and build contractor and Mott MacDonald was appointed by Trant to provide design services; it was also appointed by Trant to be the BIM coordinator being (in that particular case) the organisation that controlled access to the CDE. A fee dispute arose between Trant and Mott MacDonald and, as a result of that dispute, Mott MacDonald suspended its performance and in doing so blocked Trant's access codes to the CDE, leaving Trant and the rest of the project team unable to access the design materials. Pending resolution of the substantive dispute, Trant made an application to court for an interim injunction requiring Mott MacDonald to provide access to the CDE to Trant and to the others involved in the project. Under the circumstances of this case, the court ordered that Mott MacDonald should restore access to the relevant design materials, subject to Trant making a payment into court.

Following this decision, fresh debate has begun in the BIM community as to who is best placed to take on the responsibility of hosting the CDE. Despite the potential issues outlined above, there is a strong school of thought which says that the appointing party should host CDE as this would allow the appointing party to have full control over access to the data for the entire lifecycle of the built asset. In this way, if a dispute arose between the appointing party and a participant in the BIM process (particularly a party that was hosting the CDE), this would make it less likely that the dispute would have a detrimental effect on the access to the CDE. All that could happen in such a case is that the appointing party could either revoke the access of the party in dispute (subject to the terms of the appointment etc.) or the party in dispute could remove or stop uploading their own data to the CDE.

However, despite the common sense of this suggestion, it is obviously still the case that many appointing parties would not have the technical knowledge and capability to fulfil the role of CDE host; further, they may well say that they do not want to carry that obligation in any event.

Another alternative is for the hosting and managing of the CDE to remain within the project team, but with regular extraction of data taking place which can be stored and archived electronically and locally by the appointing party. If this approach was followed, the appointing party could ensure that if access was suddenly denied or was lost entirely, it would still have relatively up-to-date data back-up, which may only need a small amount of updating to allow the project to continue to function. Of course, although this approach may make the difficulties that may be faced slightly easier to bear, there would still be the need to procure and run a new CDE going forward.

Alternatively, the responsibility for hosting the CDE could be split between the appointing party and another member of the project team. Under this approach, the design materials could be hosted entirely on the project team member's CDE up until the point where they pass into the 'published' zone of the CDE; at that point, the design becomes valuable to the appointing party and so is moved from the project team member's CDE to a CDE hosted by the appointing party.

Of course, how the CDE is to be hosted remains a project-specific decision, bearing in mind factors such as capabilities, technical knowledge, size of the project and risk allocation. All these factors must be clearly reflected in the appointment conditions and information protocols of all relevant parties. Clearly, parties need to ensure that they are adequately protected in respect of this matter and that data required for the success of the project as a whole is not allowed to be used as a bargaining tool in a bi-party dispute.

As noted above, the CDE is not just an electronic filing system, it is also a general data centre for the obtaining, storing, sharing and extraction of data and other information, in a managed and controlled way. Therefore, the CDE needs to be the subject of processes and rules to make sure that only authorised participants can access the CDE and that, when accessing the CDE, those participants use only the current version of a data file or a model and only add to, delete or amend the data or the model accessed in the way that is permitted by the CDE rules and procedures. Usually, the participants permitted to access the CDE are provided with codes to allow access to the CDE or to parts of the CDE, as applicable.

The organisation hosting and managing the CDE is essentially a procedural gatekeeper, policing the CDE to ensure that it follows the agreed protocols and that the data is kept secure. Therefore, for effective use of the CDE, strict discipline is required from all project team members to adhere to the agreed processes and procedures.

Obviously, data security of a CDE is a major issue and concern. There is a fine balance to be achieved between opening up relevant systems and storage to allow for collaboration while being sure that security is sufficiently locked down to avoid the potential of damage – be that unintentional or malicious. Whichever BIM platform is used and irrespective of where it is hosted, all project participants need to be aware of the potential cyber-security issues. The most obvious of these is the loss of or the corruption of data held on the platform.

The matter of information security is dealt with in more detail in Chapter 10 of this book.

Construction operations building information exchange (COBie)

Construction operations building information exchange (COBie) is a spreadsheet data format that contains digital information about a building. It is often thought of as a subset of a building model. COBie focuses on delivering building information rather than geometric modelling.

The version of COBie most often used in the UK presently is the COBie 2.4-UK version. An updated version, more in line with the BS EN ISO 19650 (BSI, 2019) standards, is currently in development.

The handover of projects can clearly be improved by the delivery of consistent data through an agreed and standardised process such as COBie (or something similar), and several computerised maintenance management systems have implemented the direct import of COBie information.

The use of COBie simplifies the work required to capture and record project handover data. Its approach is to enter the data as it is created during design, construction and commissioning, and there is therefore no duplication of effort when this information is passed over at the handover stage.

Clause 5.6 of BS EN ISO 19650-2 (BSI, 2018b) specifically addresses the collaborative production of information, and the BS EN ISO 19650 series emphasises collaborative work in producing information. The BS EN ISO 19650 series recognises that parties involved in a project do not always maintain a complete record of electronic data throughout the project lifecycle, but makes clear that access to relevant information should be provided for the duration of a party's liability under their appointment.

5.7. The information delivery model

A BIM information delivery model is a digital representation of a built asset across its lifecycle, integrating structured, multi-disciplinary data from planning and design to construction and operations. It defines how objects, classifications, properties and values are delivered and are to be exchanged. It contains a set of structured and/or unstructured information containing any combination of geometrical information, alphanumerical information, both graphical as well as non-graphical (databases, spreadsheets etc.) and documentation. It provides a shared knowledge resource for information about a built asset, from initial conception to demolition, and BIM objects, and offers detailed information about parts, materials and systems for construction managers.

At an early stage, appointed parties need to check the availability of reference information and shared resources, generate information, undertake quality assurance checks, review information and approve it for sharing, and submit the information delivery model for authorisation and acceptance by the lead appointed party.

The lead appointed party then needs to carry out an information delivery model review and confirm acceptance (or not, as the case may be).

The appointed parties' separate information models are then federated into one information delivery model by the lead appointed party, which is sent to the appointing party for the appointing party's review against the acceptance criteria. The appointing party will then either approve or reject the federated information model as a BIM deliverable. A published information delivery model is one which has been accepted as a deliverable by the appointing party. It is normally a 'read-only' version and is accessible only by the project team.

The information model approval process as outlined above occurs for each information delivery milestone (data drop), as specified by the appointing party's exchange information requirements.

5.8. Close out: lessons learned

Once all delivery milestones have been met and all appointments have been completed, the information model is aggregated and archived to enable the ongoing operation and maintenance of the built asset. This ensures there is a definitive final version of the project information model available in case it needs to be referred to by the appointing party after the project has been completed.

This archive stage is essential as a future knowledge base, as well as to meet regulatory and legal requirements. Archived information includes the project history (file change history) and other metadata.

Archived information is normally read-only information and is only visible to the appointing party and the CDE administrator.

Asset information model (AIM)

The ultimate output of the information management process is the creation of an asset information model (AIM), which provides all the information required to operate and maintain the built asset. At the end of the project, some components of the project information model may be transferred to the asset information model, while unnecessary/surplus data is archived. The asset information model can include graphical models, non-graphical data and all necessary documentation for the ongoing maintenance, operation and management of the asset.

Presently, most projects require the handover of operation and maintenance manuals (O&M manuals) containing paper documents detailing equipment lists, product data sheets, warranties, spare part lists, preventive maintenance schedules and other information. This information is essential to support the operation, maintenance and management of the built asset by the client, end-user or the facilities manager.

The asset information model should contain all the specifications, O&M manuals and warranty information that would be required for future maintenance and/or repair of the built asset. The production of an asset information model helps to eliminate the problems that can currently be experienced if the O&M manuals (after they have been produced) have been misplaced by the appointing party, client or end-user (which occurs far too often presently) or are kept at a remote or different location from where they need to be. In any event, gathering the required information at the end of the construction phase, which presently is the standard practice, is very time consuming and expensive, particularly as often most of the information has to be gathered and/or recreated from information that was more readily available and/or that had previously been created at an earlier stage of the construction process.

Capture lessons learned

An important, but often overlooked, step of the entire BIM process is to review and capture lessons learned and communicate these with the BIM participants to enable the continuous improvement of the delivery and operation of the BIM process as applied to built assets. This exercise should be undertaken by the appointing party with assistance from the lead appointed party.

As the project nears close-out and the project information model is completed and the asset information model is being produced, the appointing party will need to archive the information containers in the CDE and the final model(s) and will need to capture lessons learned with each lead appointed party.

The opportunity to capture, store and disseminate lessons learned from one project to the next (and from one appointment to the next) supports continuous improvement of the briefing, delivery and operational outcomes of future BIM projects. Although this is noted as a project close-out activity, this does not only need to be carried out at the conclusion of all activities as the interim capturing of lessons learned might also improve processes within the duration of any particular appointment.

It is of paramount importance that sufficient time and resources are allowed to identify, record and understand the implications of the lessons learned. A robust process will ensure information collected is appropriately structured and in a consistent format to enable analysis, storage and onward utilisation.

The capture of lessons learned is also an activity that is referenced in the BS 8536 (BSI, 2022b) series (which currently remains within the UK BIM framework) as part of the soft landings approach. Soft landings provides a structured methodology whereby lessons learned during the BIM process can be captured. The key objective of capturing lessons learned is to support future projects from briefing through to the operational stages. Therefore, consideration should be made to engage with participants across the project lifecycle to best inform the lessons learned and in doing so assist in operating the BIM process on future projects.

Through a soft landings approach, the implementation of suitably developed post-project evaluations and post-occupancy evaluations provide the platform on which project lessons can be captured from a variety of project participants (e.g. design team, facilities managers, contractors, end-users etc).

To ensure that it operates effectively, it is important that obligations to actively participate within the lessons learned process are clearly embedded within the various participants' appointments. The project participants need to be engaged and need to be encouraged to contribute to the lessons learned and, in doing so, consider what was done well, what could have been done better, and areas for further discussion/review. Both quantitative and qualitative information should be collected (as appropriate) and suitable provisions for the analysis and dissemination of this information should also be considered within the organisations concerned.

Typical assessment criteria that may be considered within a lessons learned exercise include, but are not limited to, the following.

- Assess whether the project/investment delivered the required outcomes.
- Assess any pre-determined performance metrics.
- Did the project delivery meet the required budget and programme?
- Did the procurement process satisfy all parties?

- Did the information management process deliver its required outcomes?
- What is the built asset user's feedback?
- What are the final commercial costs for the project for benchmark purposes?
- Does the operational asset perform as designed?
- What were the social benefits/values delivered by the project?
- What was the carbon impact of the built asset?

Linked in with the above is the 'golden thread' which is a requirement arising (in the UK) from the Building Safety Act 2022 (HMG, 2022). The golden thread is both the information that allows one to understand a built asset and the steps that are needed to keep both the building and people safe, now and in the future.

The golden thread needs to hold the information that those responsible for the building require to show that the built asset was compliant with applicable building regulations during its construction and to provide evidence of meeting the requirements of the new building control route throughout the design and construction and refurbishment of the built asset. Also, it needs to provide information to allow for the identification, understanding, managing and mitigation of building safety risks in order to prevent or reduce the severity of the consequences of fire spread or structural collapse throughout the lifecycle of the built asset. The information stored in the golden thread process needs to be reviewed and managed so that the information retained achieves, at all times, the stated purposes. The golden thread covers both the information, the documents and the information management processes (or steps) used to support building safety, and the information should be stored as structured digital information.

Many people may need to access the golden thread to update and share golden thread information throughout a built asset's lifecycle, including but not limited to building managers, architects, contractors and many others. Information from the golden thread may also need to be shared by the accountable person (appointed under the (UK) Building Safety Act 2022 (HMG, 2022)), together with other relevant people including residents and emergency responders.

5.9. Summary

- The appointing party needs to appoint one or more lead appointed parties.
- The lead appointed parties need to appoint one or more appointed parties.
- An invitation to tender needs to be produced and that should include the project BIM brief, the information protocol, the exchange information requirements, the organisational information requirements, the project information requirements, the asset information requirements and a BIM execution plan template.
- The tender response to the invitation to tender should include a pre-appointment BIM execution plan and a risk register.
- A mobilisation plan and a project implementation plan need to be produced.
- Appointments can then be made incorporating the post-appointment BIM execution plan, the updated responsibility matrix, the applicable task information plans and the master information delivery plan.

- During the production phase, the collaborative production of information is collected, stored and distributed from a database called a common data environment (CDE); certain elements of the data may be contained on a spreadsheet (commonly on COBie (the construction operations building information exchange)).
- The information delivery model needs to be created.
- An asset information model (developed from the information delivery model) needs to be produced.
- Lessons then need to be learned and the 'golden thread' of information needs to be developed.

emerald PUBLISHING ice

Peter Barnes
ISBN 978-1-83608-515-7
https://doi.org/10.1108/978-1-83608-512-620251007
Published under exclusive licence by Emerald Publishing Limited

Chapter 6
The legal challenges raised by BIM

6.1. General

The use of the BIM process in both the public and private sectors is increasing, albeit with differing levels of application. One of the many perceived benefits of the BIM process is that it will reduce the risk and cost of projects, but this will only be realised if contract documentation is adequately set up to allow the BIM process to work as it should do, and to thereby avoid disputes.

The BIM process represents what is in many ways a unique intertwining of legal and technical issues and this may have led lawyers in the past to think of an information protocol (previously known as a BIM protocol) (i.e. a document that can be used to incorporate BIM into appointments/contracts) and its associated documents as essentially technical sections of the contract that should largely be left to the professional team (similar, perhaps, to a more traditional employer's requirements document). However, this, of course, is not the case and a failure to properly address the contractual issues raised by the use of the BIM process would simply be storing up problems for the future.

Currently there is no market standard position on the approach to the BIM process and, as BIM continues to develop, it is important to have an awareness of contract amendments and to consider on a project-by-project basis whether any contract amendments are required, which largely depends on the level of the BIM application, the BIM documentation and how familiar the appointing party (often the client/end-user), the design team, the contractor and the subcontractors are with the process.

The legal profession, who are required to draft a clear contractual framework within which BIM operates, is often perceived as having little grasp of the actual working of BIM projects and how obligations relating to the use of BIM should be integrated into the contract/project documentation. To add to this, the full implementation and undertaking of the BIM process on a project requires not just an adjustment to the contractual terms and the proper incorporation of technical documentation but should also entail a change in the way the industry works. This is, of course, a significant issue in itself. Clearly, the proper adoption of the BIM process on a project should lead to more collaborative working, early contractor involvement and even, if a whole-life approach is taken to the use of BIM, early involvement of the facilities maintenance contractor in the development of the yet to be constructed built asset, given that BIM is applicable to the entire lifecycle of a built asset. The legal framework going forward will have to recognise and cater for this new approach brought in by BIM.

It is therefore important for lawyers to understand how the BIM process works, the functions of the exchange information requirements, the BIM execution plan and the information protocol, and the underlying technical features of the use of BIM models.

It is fair to say that, in its early days, the use of BIM was largely driven by the UK Government and was further pushed forward by the involvement of software providers; it is possibly because of these factors that detailed legal considerations and implications have lagged further behind than they should have done.

To counter this to some degree, BIM4Legal, now known as IM4Legal, is an initiative that supports the understanding and knowledge sharing of information management and BIM among legal professionals and those in the construction and infrastructure industries who instruct them. This initiative encourages and facilitates best practice in the use of the BIM process, particularly in the context of legal issues, contract terms and dispute resolution. Its stated aim is to serve as a neutral forum for construction lawyers, project managers and industry participants to network, gain experience and share knowledge of the BIM process. The aim of the initiative is to provide better legal and contractual support in respect of BIM, thereby reducing risk and issues within the industry.

Because the BIM process enables and stimulates a collaborative working platform for all participants in a construction project, traditional contracts, which were structured to administer largely fragmented working conventions and practices, are generally lacking when dealing with the collaborative environment established and required by the BIM process. The development of specific contractual provisions and guidelines for BIM-enabled projects will become necessary to overcome the deficiencies in existing conventional contracts.

6.2. UK BIM Alliance

The UK BIM Alliance (now known as 'Nima') was launched in 2016 as a cross-sector alliance spanning the built environment to advocate for the adoption of BIM in the UK and for wider digital transformation. The UK BIM Alliance's (Nima's) aim is to ensure that BIM becomes mainstream while transforming and future-proofing the way the construction and infrastructure industries work. The UK BIM Alliance's (Nima's) objectives include providing independent and trusted leadership, mobilising communities, supporting practical implementation in businesses of all sizes and developing a common understanding and approach towards BIM. Its activities include collaborative projects to foster digital transformation, and resources and events to inform and guide all relevant professional communities. The UK BIM Alliance (Nima) is a partner of the UK BIM Framework, offering resources and guidance for implementing BIM across the UK built environment.

The UK BIM Framework (which followed on from the now defunct Centre for Digital Built Britain which ceased operations at the end of September 2022) itself develops standards, guidance and resources to support individuals and organisations in the UK to understand the fundamental principles of information management using the BIM process and to guide and support its implementation.

The Global BIM Network, established in March 2021, connects international public sector representatives and multi-national organisations with the aim of advancing the digitalisation of the global built environment and sharing the resulting benefits.

The Global BIM Network is the natural development of the increased level of international collaboration between governments and multi-lateral organisations on the digitalisation agenda in the construction industry and built environment. The resulting partnerships have stimulated national BIM programmes and regional initiatives that promote the use of the BIM process across the public and private sectors, aligned with a common language and common benefits.

6.3. Legal checklist

If BIM is used by a party 'internally' for a project – that is, where members of the same company work together under the auspices of BIM but not with other external parties – this is unlikely to require any changes to the appointment conditions and simple common BIM standards would be adequate for use.

However, where BIM involves third parties, which would apply in a vast majority of cases, consideration needs to be given as to how to incorporate BIM into the relevant appointments/contracts.

Putting to one side the position that, at some time in the future, and because of the intended collaborative nature of BIM, it is possible that no individual appointments will be needed at all as they will be replaced by 'integrated practice' and collaboration between the parties (we are not there yet), the most likely approaches at the moment still rely on bi-party (or possibly tri-party) appointments, with BIM incorporated either

- by the addition of an all-encompassing information protocol (previously referred to as a BIM protocol)
- by amending existing standard form contracts
- by way of new contract forms.

Irrespective of what approach is followed, the following checklist for the appointments relating to the use of the BIM process will be a useful starting point.

Parties and participants

- The appointing party needs to be stated.
- The BIM information manager needs to be stated.
- As applicable – the lead appointed party needs to be stated.
- As applicable – the appointed parties need to be stated.

Obligations of the parties and participants

- The obligations of the parties need to be stated.
- The works need to be described and stated.

- The appointment documents need to oblige the various parties to comply with the directions of the BIM manager/BIM coordinator but this should not extend to the BIM manager/BIM coordinator having any design/construction powers or responsibilities.
- Clauses requiring compliance with local, national and international regulations governing the application of BIM need to be included in order to maintain legal conformity.

The contractual adaptation for BIM

- There must be accepted, clear and binding contract terms.
- In the same vein, the appointment/contract needs to be reviewed as a whole (including the BIM provisions) to check for inconsistencies and discrepancies between the appointment/contract documents.
- It needs to be established how BIM is to be incorporated into all of the contractual relationships of the various parties.
- Back-to-back BIM terms and standards across the project team are essential to ensure consistency of the process and to avoid gaps in responsibilities and liabilities. Parties may find this most easily achieved by expressly requiring the incorporation into all contracts down the supply chain of the same comprehensive information protocol (previously referred to as the BIM protocol) or other similar stand-alone document.
- If a separate information protocol is to be used, consideration needs to be given as to whether that should be included as one of the usual contract documents within the various appointments, or whether it should be a stand-alone amendment to the appointments.
- If an information protocol is used, it needs to be established whether the information protocol takes precedence over the appointment/contract if conflicting terms arise.
- A clause dealing with omissions, ambiguity, conflict or inconsistency in or between any information prepared in relation to the works in respect of the BIM process needs to be incorporated.
- Contractual reporting requirements for known or observed errors or omissions in contract documents should be included to ascertain whether they are adequate and consistent given the potentially increased pace of electronic document exchanges.
- It must be established that any information protocol that is produced does reflect the actual intentions of the parties in respect of the BIM process.
- The information protocol must include obligations on all project team members to produce and deliver models and comply with the information requirements.
- The information particulars (i.e. the project-specific terms and conditions applicable to the BIM process) need to be accurately completed.
- As collaboration is at the heart of the BIM process, collaboration therefore needs to be supported by clear and accepted collaborative or 'mutual trust' contractual terms. An optional or non-binding 'collaborative' understanding gives rise to a real risk that the collaborative promises will simply fall by the wayside if a dispute arises.
- Serious consideration needs to be given to the incorporation of binding good faith provisions.
- Definitions need to be provided. There should be consistent definitions and terminology between all the various documents, noting, in particular, that published standard

information protocols are likely to use definitions and terminology that may be different from other appointment documents.

- Clauses relating to the requirement for training and education of project participants regarding legal and contractual obligations in the context of the BIM process need to be incorporated.
- The design liability obligation/level of duty and skill and care must be clearly stated.
- Clauses relating to the termination of the appointment need to be incorporated into the appointment. The consequences of termination may depend on the reason for the termination but it is likely that both parties will require some access to the CDE following the contract being terminated, and this needs to be catered for.

Standards

- Clauses relating to adherence to stated or established industry standards governing BIM practices (e.g. BS EN ISO 19650 (BSI, 2019)) need to be included.
- To promote standardisation, it may be sensible to specify, where possible, which BIM library/libraries will be used, such as the NBS National BIM Library (now known as the NBS BIM Library).
- An agreed data dictionary (such as the building SMART data dictionary) should be stated in order to catalogue standardised meanings of the data being exchanged in the models, for the parties' common understanding.

BIM execution and delivery plans

- Details of the BIM execution plan need to be stated.
- The master information delivery plan needs to be provided.
- The task information delivery plans need to be provided.
- The responsibility matrix (both high-level and detailed) needs to be provided.
- The mobilisation plan needs to be provided.
- The required BIM deliverables/outputs must be identified.
- The standard of care in respect of the deliverables should be stated – that is, 'reasonable endeavours', 'reasonable skill and care' and so on.
- Specifying the priority of BIM deliverables and the models may also be sensible as terms may provide for both 2D drawings and 3D models to be deliverables.
- The contract documents should place an obligation on all parties to comply with the BIM process and timetable as this will aid consistency of the process and facilitate a collaborative mindset, as all parties will feel they are 'in it together'.

Exchange information requirements, documentation and record keeping

- The exchange information requirements need to be set out.
- The same exchange information requirements (amended only as necessary) for all appointments are preferred as this facilitates consistency and helps avoid misunderstanding or miscommunication (which could in turn lead to disputes).
- Establish protocols for the retention and accessibility of project-related documents for legal purposes.

- Methods must be established for maintaining version control of electronic documents, including a depository of record copies of transmitted and received electronic documents.
- An agreed BIM workflow should be incorporated. This assists the effective implementation of BIM by helping standardisation of the entire BIM process. The five golden rules of a BIM workflow are accessibility, work sharing, applicability, reliability and efficiency.
- The timetable for production and delivery of the models and other BIM data to ensure the scheduled programme for the works is met must be stated.
- An express obligation to comply with the BIM programme, information plan and related time-sensitive matters is a vital requirement.

Models

- State whether the models are to be contract documents. As models will be evolving and changing over the course of the design period, it is highly questionable whether the models should be contract documents.
- If the models are not to be contract documents, it needs to be stated how the models are to be incorporated into the appointment.
- State what models are required to be created, their relationship and the relationship between the model(s) and the 2D drawings. It is also important to define the position of 2D drawings which will normally still (presently) be required for obtaining building permits, and the level of information/reliance to be placed on them; this will be important especially in terms of the level of detail in a project.
- Where there are multiple models, the parties should indicate which model takes priority in the event of conflict in the information contained in each of the models. Procedures and safeguards for clash detection and checking should be considered to minimise the risk of an error in the data or models which will be copied and relied on 'infecting' all the information flowing from it.
- Who owns the models (and the information contained within them)? In most projects, ownership of the BIM model is likely to be retained by the owner of the building (normally the appointing party). However, ownership of the data contained within the BIM model itself is a separate issue. Such data is likely to be wide ranging and contributions will come from a variety of different participants – for example, there are likely to be design data, cost data, energy data, facilities management data, design processes, tables, databases and graphical information.
- Note that a dynamic or parametric model (i.e. a self-generating/self-evolving model) creates very challenging legal issues around ownership and preservation.
- State what design information needs to be contained in the models, including the level of detail and the level of information need (and at what stages that information is needed).
- It needs to be made clear who is responsible for updating and preserving the models (and the information contained within them) to make sure that they comply with the exchange information requirements.
- It needs to be stated what BIM deliverables are required from the various BIM models.
- The parties would need to agree on the content and format of each model and the required standards to be employed in creating them.

- Consideration should be made with regard to excluding statutory implied terms such as fitness for purpose under the Sale of Goods Act 1979 or Supply of Goods and Services Act 1982 insofar as models are concerned (at least until the law is clarified as to whether those statutes apply to BIM models).
- A no liability clause arising from any modification or amendment to the information model should be considered.
- A procedure needs to be established which sets out how design changes and variations will be registered and/or notified to other parties once implemented into the models, thereby avoiding parties relying on out-of-date information.
- Clauses relating to the standard of care required when providing information for the model (normally set at the level of reasonable skill and care) need to be stated.

Software

- The software to be used (which, ideally, needs to be inter-operational) needs to be stated.
- It needs to be made clear who is responsible for obtaining software licenses.
- It needs to be stated who is liable (or where the liability lies) in respect of the suitability and integrity of the software selected.
- It needs to be made clear who is liable (or where the liability lies) for software failures (e.g. power interruptions, viruses and physical damage). Software manufacturers are normally protected by blanket limitation of liability clauses that makes it inappropriate to attempt to transfer the risk of BIM errors caused by software to the manufacturers. This risk should therefore be dealt with in the appointment/contract documents.
- Clauses relating to the acceptable form of testing and commissioning and/or assessment and inspection of models and data at handover to confirm integrity and accuracy need to be incorporated.

Data ownership and liability

- There needs to be clarity regarding the legal responsibilities in relation to the ownership of data generated within the BIM process.
- Liabilities associated with the accuracy and application of BIM-derived information need to be defined.
- Requirements need to be put in place for collaborative information sharing, which impose reasonable limitations on the use and reliance of data and models supplied (e.g. excluding liability for changes by third parties and any use for other projects).
- No liability for any corruption or unintended amendment, modification or alteration of any data and/or software should be dealt with.
- Clauses should be added dealing with database rights.
- The method of keeping and maintaining digital information about an asset (e.g. COBie) should be specified.
- Details of the common data environment (CDE) should be set out.
- Clarification of who hosts the CDE and who has access to the CDE needs to be provided.
- Clauses dealing with the security, integrity and preservation of the information archived in the CDE need to be included.

- Clauses need to be included relating to continued interoperability when errors or loss of integrity occur when different systems exchange information which do not relate to one another.
- Specified common standards and formats for data exchange need to be provided as this will reduce interoperability issues further.
- Although the above dangers could be reduced by appropriate back-up strategies, there are risks involved with hosting data; BIM models, like all digital data, are susceptible to data loss, and even small data losses can require significant effort of recovery or replacement. Therefore, if a party is hosting the information, it must take adequate steps to protect, and insure, against data loss, or it will face possible liability for the ensuing losses (which could be very significant in size). A design firm can purchase 'valuable papers' coverage that provides catastrophic loss protection, but this will not necessarily cover losses to other collaborative users. These aspects all need to be covered by the contract/appointment terms.

Data confidentiality and security

- Measures to safeguard the confidentiality and security of BIM data must be implemented.
- There may be a need for confidentiality clauses or separate confidentiality or non-disclosure agreements where the design or other information revealed and/or shared by the construction team members and/or the employer is commercially sensitive.
- A definition of protocols for data access, sharing and storage to prevent unauthorised use needs to be provided.
- The security management plan needs to be provided and an ongoing liability for data security needs to be dealt with.
- Clauses need to be inserted requiring the observation of the appropriate data protection regulations (including GDPR (General Data Protection Regulation) in the UK).
- Hosting, storage and security: details must be provided regarding where the models are to be hosted and stored, and what security measures will be taken against data degradation, corruption and hacking (a particular risk for commercially sensitive projects). It may be sensible to specify the format and location of back-ups of the models and related data as well in the contract/appointment terms.

Risk management

- A clause in respect of the need to develop risk management strategies to mitigate potential legal and financial risks associated with BIM projects should be added in.
- Within the contracts/appointments, there should be consistent risk allocation at all levels and/or variations in the levels of risk associated with each party's responsibilities and rights; any limitations on risks must be clearly identified.
- A risk register needs to be provided.
- The risk allocation in an information protocol must be checked (and/or amended, as necessary) to ensure that it reflects the risk allocation agreed between the parties.

Intellectual property protection

- Concerns related to the protection of intellectual property within the BIM framework need to be addressed.
- Protocols for the use, sharing and protection of proprietary information need to be established.
- Clauses dealing with copyright, moral rights, rights in design and database rights need to be incorporated, noting that copyright and intellectual property provisions are an implicit form of ring-fencing information in the agreement, and a single island of information in this way is unlikely to be in the project's best interest or in the collective interest of the entire BIM team.
- The issue of non-exclusive, royalty-free and irrevocable copyright licence/sub-licences needs to be addressed.
- Clauses relating to indemnities for infringement of intellectual property (IP) rights need to be added.

Insurance

- Add clauses dealing with insurance requirements related to BIM implementation.
- State the insurance or assurance required regarding the integrity of data in the models and BIM deliverables from all participants – including the BIM information manager and the BIM coordinator.
- If a failure occurs, state what insurance, if any, will respond to the economic losses, noting that coverage under the designer's professional liability policy is problematic, and the designer's commercial general liability policy will not normally cover pure economic losses. The difficulty in characterising and insuring against this type of loss underscores the need for comprehensive risk allocation and waivers among all BIM participants.
- Where a project-wide insurance is available this should be adopted.
- All project participants should be required to waive their rights to consequential damages and limit liability to direct losses connected with the project.

Close-out and soft landings (and the 'golden thread')

- Terms imposing obligations on the construction team for soft landings (and compliance with the 'golden thread' pursuant to the Building Safety Act 2022 (HMG, 2022)) should (must in some cases) be incorporated and may in practice lead to fewer claims post-completion, given that issues and problems immediately arising post-completion will be resolved, and the end-user will have a better understanding of the property in general.
- Terms relating to the handover of the asset to facilities management should be incorporated.
- A requirement for data preservation (particularly in respect of the limitation period requirements) needs to be incorporated. Although it is possible to save a model onto electronic media, this does not guarantee that the saved model will be usable (because of software/hardware changes and interoperational issues). It must be noted that in certain circumstances, the limitation periods under construction contracts may be very significantly extended pursuant to the provisions of the Building Safety Act 2022 (HMG, 2022).

Dispute resolution mechanisms

- Establish clear dispute resolution mechanisms specific to BIM projects.
- Define the procedures for resolving conflicts arising from contractual disputes or disagreements in the BIM context.
- Consider a tiered and/or collaborative dispute resolution process to promote collaboration. It should be noted that any tiered and/or collaborative dispute resolution process cannot (in the UK) remove the parties' right to statutory adjudication at any time, although promoting a culture of collaboration should assist in minimising any party's desire to resort to such a process.

6.4. Summary

- BIM cannot be simply considered as being about technical matters – the legal and contractual issues are equally important.
- The UK BIM Alliance (now known as 'Nima') was launched in 2016 as a cross-sector alliance spanning the built environment to advocate for the adoption of BIM in the UK and wider digital transformation.
- Currently there is no standard market-led approach in respect of the incorporation of BIM into appointments/contracts.
- Irrespective of how BIM is incorporated into appointments/contracts, it is important first that a legal checklist of issues that are relevant to a particular project is produced.

Peter Barnes
ISBN 978-1-83608-515-7
https://doi.org/10.1108/978-1-83608-512-620251008
Published under exclusive licence by Emerald Publishing Limited

Chapter 7
The contractual framework for incorporating BIM

7.1. How is BIM incorporated into contracts?

It is clear that the obligations and requirements associated with the BIM process should be set out in some way or another in the contract or appointment documents but there is no standard approach as to how those obligations and requirements should sit within the contract documents.

Presently, the majority of the construction and civil engineering industries still use standard form contracts (normally with amendments) on most projects, and, by and large, those standard form contracts envisage a simple insertion of an all-encompassing information protocol (previously known as a BIM protocol) containing all necessary BIM-related obligations and requirements.

However, given how BIM has the capacity to change the relationship between participants in the construction and civil engineering industries, and given that a requirement of the BIM process is for there to be more collaborative working, the present-day standard form contracts, even with an information protocol added in, are not an ideal solution. It seems inevitable that new forms of contracts will need to be developed (this matter is dealt with in more detail in Chapter 8 of this book).

Despite the above future possibility, at the present time, an information protocol is still the most commonly used method of dealing with the contractual aspects of the BIM process.

One main benefit of a separate information protocol is that (with amendments as necessary) the same information protocol can be used for all BIM participants (including main contracts, consultant appointments and sub-contracts), thereby creating a great deal of consistency of approach across the separate appointment documents between the various BIM participants.

Also, a separate information protocol can deal with the more technical aspects of the BIM process, while the relevant terms and conditions of the contract are simply updated to incorporate and weave in the necessary provisions dealing with obligations, liabilities, ownership, intellectual property and so on in respect of the BIM process. This approach allows parties to largely retain the contracts they are accustomed to while adopting BIM through a separate information protocol.

However, something to be aware of is that, where an information protocol is used, there may be divergences or discrepancies between the information protocol and other contract documents.

While every effort should be made to avoid any such divergences or discrepancies, where such do occur, it needs to be clearly stated whether or not the information protocol has priority over other contract documents in respect of BIM issues; this matter is normally dealt with by way of a contractual priority clause.

This is a very important point because if, for example, the information protocol was included as part of the employer's requirement in a JCT Standard Building Contract (JCT SBC) (JCT, 2024b), clause 1.3 of that contract would apply, and that clause gives priority to the terms of the JCT SBC over any other contract document. Therefore, in the situation as outlined above, if there was a fundamental discrepancy between the conditions of a JCT SBC and the employer's requirements document (incorporating the information protocol), the JCT conditions would take priority over the employer's requirements (incorporating the information protocol), and this would obviously either lessen or nullify the effect of the information protocol. The above may be what is required, and if so that is of course for the parties to decide, but it is an important point to keep in mind when putting contract documents together.

The matter of the priority of contract terms in respect of the JCT SBC as outlined above was dealt with and confirmed in the case of *Fenice Investments Inc.* v. *Jerram Falkus Construction Ltd* (2009) where the court decided that clause 1.3 of the JCT SBC had priority over the employer's requirements, the contractor's proposals and the contract sum analysis and could not be overridden or modified by anything contained in those documents.

As the information protocol is put in place to deal with the contractual issues relating to BIM, rather than the more general contractual issues covered in the other contract documents, it is important to provide for the priority of the information protocol over other contract documents on BIM issues to avoid the effect of the information protocol potentially being diminished or expunged entirely.

7.2. The information protocol

The information protocol (previously referred to as the BIM protocol) is a legal document which defines the parties' rights, roles and duties in compliance with the project's requirements and standards relating to the BIM process.

The information protocol is initially sent by the appointing party to the prospective lead appointed parties and is executed by the selected lead appointed party on the conclusion of a successful tender response. The information protocol formalises the appointment between appointing party and lead appointed party.

The relationship between the appointing party and a lead appointed party (by way of the information protocol) is set out in Figure 5.3 of Chapter 5.

As noted above, the information protocol, or an amended version of this, is subsequently used by the respective lead appointed parties when making appointments with the relevant appointed parties.

Given the importance of the information protocol, both it and its appendices need to be made available to all relevant parties pre-appointment so they understand the obligations and liabilities in respect of the BIM process that they will be placed under. Any changes to the information protocol and/or to its appendices need to be treated as variations to the appointments/contracts, following suitable change control procedures.

One of the main purposes of an information protocol is to maximise production efficiency through adopting a coordinated and consistent approach to working in BIM throughout the entire supply chain. It is also used to define the standards and best practices that ensure delivery of high-quality data and uniform drawing output across an entire project.

A project-specific information protocol helps to ensure that all project stakeholders, including architects, engineers, contractors, subcontractors and facility managers, work together efficiently and effectively, leveraging the full potential of BIM technology to deliver better designed, constructed and managed built assets.

The particular information protocols used on a project can vary depending on regional standards, project complexity and the appointing party's requirements, but generally include a set of guidelines, standards and procedures that help define how BIM is implemented and managed, and often address issues related to data exchange, information sharing, project roles and responsibilities, and compliance with standards.

An essential element of the information protocol is to ensure that digital BIM files are structured correctly, thereby enabling efficient data sharing while working in a collaborative environment across multi-disciplinary teams, both internally and in external BIM environments.

To be effective, every BIM project must have its own information protocol and, because of the unique nature of construction projects, each information protocol needs to be tailored to suit a particular project.

An information protocol sets out the ways in which each project has its own unique organisation and structure in respect of the BIM process. It provides the project team (or at least those appointed under it) with a road map to understanding the aims and objectives of the BIM process for a particular project; it sets out rules to be followed and, among other things, also establishes what models are required and how those models are to be assembled. Also, an information protocol applicable across the entire BIM team provides new team members with the means to comprehend and participate more easily and with more certainty in what is often, by necessity, a complex structure and process.

As outlined above, an information protocol is not, generally, intended to restructure contractual relationships or stand as a substitute for a complete building project agreement/contract; it is simply intended to be an addendum to the building project agreement/contract and to the various consultants' and other designers' appointments.

A typical information protocol document would contain

- an introduction to the project
- an overview of what BIM will be used for on the project

- the place of the information protocol in the priority of the contract documents
- a reference to any other documentation that should be taken into consideration
- details of the BIM information manager
- details of the BIM coordinator
- the exchange information requirements
- an organogram that shows in a simple way how all the relevant parties contribute to the BIM process, their obligations, roles and responsibilities, and the delivery timescales required.
- the relevant BIM execution plans
- details of the project information model, the future asset information model and details of any other models required
- the levels of detail, level of geometry and level of information need for the various BIM models
- details of the 'data drop' requirements at key stages during the development of the project
- details of the common data environment
- details of the extent to which there will be collaborative working and the use of standard specifications, BIM standards and so on
- definitions of the terminology used in the information protocol document
- details of the software to be used
- a matrix showing the different types of software used and the methodology by which file exchange between each type of software needs to be provided
- details of the handover procedure from one participant to another
- a model assembly diagram which allows each team member to understand how the BIM models are to be arranged
- a clearly set out project file structure and security protocols
- details of the data security and back-up. This includes when record copies of the BIM information and data are to be retained and to whom they are to be issued. The level of security will vary from project to project, with military or government projects likely to have extensive security measures while other projects may require lesser measures. However, it is essential in all collaborative BIM projects that no one contributor to the model is able to alter the work carried out by another contributor without proper and valid permissions having been obtained
- the key common data for coordination of setting out
- whether additional data (e.g. cost, programme etc.) are to be linked to elements and models
- details of how and where new objects or elements are to be created and by whom
- how elements are to be classified and the library structure for the project as a whole
- who has the right to add or remove layers to the various models and what the procedure for adding or removing the layers is
- the method by which changes to the model are managed and recorded. As the BIM process enables consultants to work concurrently, rather than following the more traditional sequential workflow, the setting out of a formal method by which consultants request changes (and permissions are given) will allow changes to be tracked and authorised and the responsibility for any late changes (which may, for example, affect the project schedule) to be noted

- a facility for reviews of the information protocol to be made, and for amendments to be made and notified to all relevant parties
- a definition of who has use of the BIM model(s), who can amend data once it has been incorporated into the BIM model(s), and who can view the BIM model(s) but cannot actually amend the BIM model(s)
- details of how issues regarding copyright and intellectual property rights are dealt with and how licences for these can be granted to other participants for permitted purposes
- any limitations on liability associated with the BIM model(s), and the levels of reliance that one party may place on another party's data and/or model.

If the information protocol states that there is no warranty from the project team members regarding the integrity of the electronic data they provide, this could potentially undermine the usefulness of the model. Each project team member should be able to take responsibility for the electronic data provided. In some cases, the appointing party is liable for faults in information provided by information providers other than the contractor – this would be a problem if under a particular appointment the contractor is meant to be responsible for managing the model, and this also muddies the waters on liability where the contractor (in all probability a lead appointed party) is responsible for some or all of the information providers (e.g. subcontractors and subconsultants (appointed parties)). There may also be other general clauses limiting the contractor's liability elsewhere in the contract, which need to be looked out for.

7.3. The information protocol for use with BS EN ISO 19650-2

As noted earlier in this book, the BS EN ISO 19650 (BSI, 2019) series is an internationally recognised set of standards for managing information throughout the lifecycle of a built asset using the BIM process. That series includes several parts, with BS EN ISO 19650-2 (BSI, 2018b) focusing on the delivery phase of built assets operating the BIM process.

An information protocol under BS EN ISO 19650-2 (BSI, 2018b) is a document that outlines the rights, roles and responsibilities of the parties involved to ensure compliance with the required standard. Such an information protocol is essential for integrating information management into project appointments and contracts and for ensuring that all parties adhere to the same standards and processes.

A template information protocol to support contracts which use BS EN ISO 19650-2 has been produced by the UK BIM Framework (UK BIM Framework, 2020).

In such a case, the information protocol does not operate independently; it needs to be incorporated into an appointment, whether that be a building appointment, a consultant's appointment or similar, for it to have contractual effect. The guidance includes an example of an incorporation clause which can be used to incorporate the information protocol into a contract, and a similar clause is provided below (UK BIM Framework, 2020, p. 6).

> The [Appointor] and the [Appointee] shall:
>
> 1. comply with their respective obligations set out in the Protocol [...]
> 2. have the benefit of any rights granted to them in the Protocol; and
> 3. have the benefit of any limitations or exclusions of their liability contained in the Protocol.

> The [Appointor] and the [Appointee] agree that, subject to clause 1.6 of the Protocol, this Appointment shall be amended as set out in the Protocol.
>
> In the event of conflict between this Appointment and the Protocol, the parties agree that the terms of the Protocol shall take precedence.

An alternative to the above would be for the appointment to take precedence, in which case, the final sentence of the draft clause noted above, would read

> In the event of conflict between this Appointment and the Protocol, the parties agree that the terms of the Appointment shall take precedence (UK BIM Framework, 2020, p. 6).

The template information protocol is intended for use at all levels of the supply chain; for this reason it refers to two main parties: the appointee and the appointer (which can be any two parties as relevant anywhere down the supply chain).

The template information protocol makes it clear that any obligation to deliver information management in accordance with BS EN ISO 19650 (BSI, 2019) means to perform the parties' obligations under the information protocol, as otherwise such a general obligation would be unclear.

Under the template information protocol, the parties have no greater obligation than to exercise the level of skill and care required under their respective appointments, thereby reducing the risk of the information protocol being inconsistent with professional indemnity insurance arrangements.

The template information protocol states that in the event of any conflict or inconsistency between the information protocol, the information particulars (which form part of the template information protocol) and any other documents contained in and/or forming part of the contract (or the appointment), such conflict or inconsistency shall be resolved in accordance with the contract (or the appointment), provided that if the contract/appointment does not specifically include provisions stating how such conflict or inconsistency should be resolved, the documents shall have the following order of precedence: (1) the template information protocol and (2) other documents forming part of the contract (appointment). Obviously, this order of precedence of documents will not always be appropriate for every project and should therefore be confirmed on a project-by-project basis.

The information protocol sets out the information particulars at the front of the document, which are the project-specific details that need to be completed for each project/appointment. Those particulars use the terminology of BS EN ISO 19650 (BSI, 2019) and include basic details such as

- the appointment (the contract)
- the appointed party
- the lead appointed party
- the appointor
- the works

- the level of information needed
- the exchange information requirements
- the BIM execution plan
- the master information delivery plan
- the task information delivery plan
- the risk register
- the high-level responsibility matrix
- the security management plan.

The success of any information protocol is dependent on all the required technical documents (including those noted above) being provided, with missing or incomplete documents leading to uncertainty as to the processes to be followed by the parties. The template information protocol is drafted so that the document could function even if the information particulars were not completed (which would be a strange approach to take); however, if the information particulars are left blank, the parties will be unclear about the key information management processes and documents, and this would almost inevitably create uncertainty and potential dispute in the future.

The clauses of the template information protocol are as set out below.

Clause 1 Interpretation of the terms of the information protocol

Clause 2 Coordination and resolution of conflicts

A key consideration when incorporating the information protocol into a contract is the approach to resolution of conflicts between the information protocol and the underlying appointment/contract. The information protocol does not state that any aspect of the information protocol will always take priority, so if the appointment/contract includes an order of precedence, that will always apply. This makes the information protocol a more flexible document. The example incorporation clause included in the information protocol guidance also allows parties to indicate whether the information protocol takes priority over the appointment or not.

Clause 3 Obligations of the appointing party (or the appointee more specifically)

Clause 4 Obligations of the appointee and appointer

The obligations in BS EN ISO 19650:2 (BSI, 2018b) regarding testing of methods and procedures, and the obligations to maintain capacity, are included in clauses 4.2 and 4.4, respectively. Clause 4.3 requires the lead appointed party to establish and maintain the delivery team's risk register. Clause 4.6 acknowledges that compliance with the applicable information particulars may give rise to an extension of time and/or additional costs under the appointment/contract, and there is an express, positive obligation under clause 4.7 for the parties to provide such information and assistance required by the information particulars. Clause 4 also deals with security, other appointments, information management documents and information delivery plans. The information protocol's security provisions take into account BS EN ISO 19650-5 (BSI, 2020b) and ensure consistency with its provisions and include references to the security management plan. The parties have no greater obligation under the information protocol than to exercise the level of skill and care required under the appointment, reducing the risk of the information protocol being inconsistent with professional indemnity insurance arrangements. Clauses 4.12 and 4.13 place obligations on parties regarding task information delivery plans.

Clause 5 The common data environment

The appointing party, in clause 5.3, is responsible for establishing, implementing, configuring and supporting the common data environment (CDE) solution and workflow. In clause 5.4, the information on the CDE solution is to be stored securely by the appointing party for the period specified in the information particulars. With reference to the decision in *Trant v. Mott MacDonald* (2017), the information protocol (in clause 5.5) confirms when parties should have access to the information held in the CDE solution. That clause provides that access should be provided for such period as the party is liable for its services under its appointment, recognising the reality that parties do not (and often cannot) maintain a complete record of the electronic data on a project which becomes vital in the event of an issue arising.

Clause 6 Management of information

Clause 6 deals with the management of information, requiring the appointment of appropriate individuals to undertake the information management function, obliging the appointor and the appointee (as stated in the information protocol) to carry out the information management tasks set out in the information particulars, including those listed in clauses 6.2 and 6.3. The parties can appoint others to undertake all or part of their information management functions but clause 6.4 confirms that any such delegation does not allow them to avoid their obligations under either the appointment/contract between them or under the information protocol itself.

Clause 7 Level of information need

It is ultimately for the appointing party to specify how much information it wants and to what detail. To avoid confusion, the level of information need must be as detailed as possible (in the information particulars) and should ideally include indicators so that parties can tell that they have reached the required level. This establishes the framework for specifying the extent and granularity of information to be provided in the information particulars.

Clause 8 Use of information

Clause 8.1 covers data protection laws (which would include the General Data Protection Regulation (GDPR) in the UK) arising from the processes to which the information protocol applies. Clause 8.1 is a short, flexible clause intended to keep pace with developments in this area rather than provide an overly prescriptive set of obligations. The remainder of clause 8 provides for licences to be issued by the parties, enabling the use of the material for permitted purposes, reflecting the contractual ring-fencing of liabilities and obligations in existing construction contracts. Any provisions in the appointment/contract take precedence and are modified only insofar as necessary to comply with the obligations in the information protocol, and, to the extent that there are no such provisions, the information protocol will apply.

Clause 9 Transfer of information

Clause 9 establishes a contractual obligation on the parties to deliver the information required by the information particulars. In addition, lead appointed parties are required to carry out risk assessments in relation to the preparation of the management of information and to reflect the long-term asset information management strategy. This is intended to tie this aspect of the design/delivery phase to the longer-term asset information requirements, reflecting the information flow structure set out in BS EN ISO 19650-1 (BSI, 2018a). Clause 9.4 requires the lead appointed party to assist the appointing party in capturing lessons learned during the performance of the appointee's obligations under the appointment. This is a useful provision which hopefully will prompt parties to attempt to preserve information rather than allow it to dissipate as project teams are disbanded and move on to the next job.

Clause 10 Liability

Neither party to the information protocol has liability to the other if the other party changes or modifies the model or related work for any reason other than for a contractually defined purpose.

Clause 11 Remedies: security

The information protocol's security provisions in clauses 4 and 11 have been updated to take into account BS EN ISO 19650-5 (BSI, 2020b) and to ensure consistency with its provisions. These amendments to the previous security provisions include, for example, new references to a security management plan.

Clause 12 Termination

Termination provisions are set out in clause 12.

Clause 13 Definitions

Information protocol definitions are set out in this clause.

It should be noted that the Construction Industry Council (CIC) published its *Building Information Modelling Protocol* in February 2013, with a second edition in April 2018 (CIC, 2018). This second edition is not compliant with the BS EN ISO 19650 (BSI, 2019) series and would need to be amended to make it compliant.

7.4. Standard form contracts and BIM

Determining an approach to collaboration as early as possible in a project is strongly advised, even if a range of details are left unresolved until later stages. Such direction of approach is likely to come from the appointing party (and their advisers) initially and they will inform the selection of procurement route, form of contract and the preparation of tender documentation.

The UK Government has previously recommended one of three forms of procurement when using BIM.

- Design and build: a procurement route where the main contractor designs and constructs the works (as contrasted with a more traditional arrangement where a client (appointing party) appoints consultants for design and then a contractor for construction). Design and build contracts solve the boundary problem between the various design consultants by, in effect, increasing the boundary's perimeter until it absorbs all the key participants. Thus, information sharing and reliance issues are resolved by joining the provider of design and the provider of construction to the one party. Therefore, in many ways a design and build procurement approach may be seen as being ideal for a BIM approach. However, even when using the design and build procurement approach there is still the risk of a split of liability with regard to the design functions in respect of a BIM project.

- Prime-type contracting: in this arrangement the client (appointing party) enters into a relationship with a contractor who acts as a single point of contact for a wider supply chain to deliver on a project, thereby keeping the liability under the control of one participant.
- Private finance initiative (PFI): PFI projects typically involve complex contractual arrangements. Within the integrated supply team there may be separate agreements for funding, design and construction and facilities management, and each agreement may have multiple sub-contracts.

Another form of collaborative procurement worth mentioning is partnering (or alliancing) – an approach where openness and trust is encouraged between contractual parties, establishing interdependencies that are essential for successful outcomes; this philosophy should be an ideal fit for the BIM ethos.

It must be noted that the standard form contracts, described below, focus only on the construction phase of built assets, and so far, none of the major standard form drafting bodies have produced a facilities management contract to deal with the operational or 'in use' phase of the building. As the facilities management of a built asset forms an integral part of the BIM process, this is something that needs to be urgently addressed.

The approach of the standard form drafting bodies in respect of BIM has obviously varied, and some of the approaches used are considered below.

JCT contracts

In May 2019, the Joint Contracts Tribunal (JCT) introduced a new practice note: *BIM and JCT Contracts* (JCT, 2019). JCT says that the aim of the practice note was to further the understanding of BIM-related legal and contractual issues and to suggest ways of approaching such issues in a collaborative and constructive way.

The practice note is split into two parts. Part A is a detailed commentary on the provisions in the applicable JCT contracts that are, or could be, relevant where the BIM process is implemented on a project, whereas Part B is an information protocol (BIM protocol) checklist, suggesting a non-exhaustive list of main topics which may be covered by an information protocol (BIM protocol).

In JCT contracts, the information protocol must be specifically identified in the contract particulars. The JCT does not publish its own form of information protocol and the parties must agree the form of information protocol to be used on a project during the pre-contract stage. The chosen information protocol should be reviewed to ensure that, from both a technical and contractual perspective, it does not contradict the terms of the appointment/contract or the other appointment/contract documents.

The information protocol is a 'contract document' under JCT contracts, but it is important to note that if there is any conflict between the information protocol and the contractual terms of the JCT contract itself, then (without an amendment being made to the JCT contract terms by the parties) the contractual terms prevail.

Thus, if an information protocol was adopted as part of the employer's requirements in the JCT Standard Building Contract (SBC), clause 1.3 of the *JCT Design and Build 2024* (JCT, 2024) contract, for example, would apply and this would give priority to the terms of the JCT contract over the information protocol (see *Fenice Investments Inc.* v. *Jerram Falkus Construction Ltd* (2009) where the court held that where there was a fundamental discrepancy or divergency between the conditions of a JCT contract and an employer's requirements document, in that particular case, the JCT conditions would take priority).

NEC contracts

The latest version of the NEC contract is NEC4 (NEC4, 2017); secondary option X10 has been added into the NEC4 to deal with the BIM process. NEC4 gives little flexibility in approach as it includes only the (BIM) information execution plan and the (BIM) information model.

NEC4 secondary option X10 includes: defined terms (at X10.1), collaboration (at X10.2), early warning (at X10.3), information execution plan (at X10.4), compensation events (at X10.5), use of the information model (at X10.6) and liability (at X10.7).

Other significant and relevant clauses in the main NEC4 (NEC4, 2017) contract are

- Clause 10.2 – 'The Parties, the Project Manager and the Supervisor act in a spirit of mutual trust and co-operation.'
- Clause 22.1 – 'The Client may use and copy the Contractor's design and documents prepared for the design for any purpose connected with construction, use, alteration or demolition of the works unless otherwise stated in the Scope and for other purposes as stated in the contract. The Contractor obtains from a Subcontractor equivalent rights for the Client to use material prepared by the Subcontractor.'
- Option X12 provides a multi-party collaboration clause.

In 2018 the NEC launched the NEC *Alliance Contract* (NEC, 2017) in which the client and all key members of the supply chain, called 'Partners', are engaged under a single contract. All members of the alliance have an equal voice and share in the performance of the alliance as a whole as opposed to their own individual performances. This could be a very useful basis for a BIM collaborative contract.

CIOB *Contract for Use with Complex Projects*

The little used CIOB *Contract for Use with Complex Projects* (CIOB, 2013) includes detailed provisions for the use of the BIM process, including those related to the role of the design coordination manager, provision of a CDE, software selection, suitability and integrity, archiving of each completed level of development, maintenance and updating of the model, indemnification of other design users in respect of contractor design and notification of clashes.

PPC2000

PPC2000 (*Project Partnering Contract*) (ACA, 2013) is a multi-party contract form which was amended by the addition of Appendix 10 in 2013 which makes provisions for BIM. It should be noted that PPC2000 was used on the UK Government's BIM pilot projects at Cookham Wood and North Wales Prison.

Facilities management contracts

All the above contracts focus only on the construction phase of built assets. As noted above, so far none of the main standard form drafting bodies have produced a facilities management contract to deal with the operational or 'in use' phase of the building in respect of BIM, although it is anticipated that this will follow as matters progress.

In the meantime, this lack of a facilities management contract is likely to raise issues regarding the interface between the construction contractor and the facilities management contractor and the liabilities during the interface period.

7.5. Summary

- Consideration needs to be given as to how BIM will be incorporated into contracts.
- Most standard form contracts rely on an information protocol dealing with BIM issues which is added as an addendum to the contract.
- The information protocol (previously referred to as the BIM protocol) is a legal document which defines the parties' rights, roles and duties in compliance with the project's standards relating to BIM.
- The present standard information protocol for use with international standards is the information protocol for use with BS EN ISO 19650-2 (BSI, 2018b).
- Updated facilities management contracts need to be developed to keep up with the increasing developments in BIM.

Peter Barnes
ISBN 978-1-83608-515-7
https://doi.org/10.1108/978-1-83608-512-620251009
Published under exclusive licence by Emerald Publishing Limited

Chapter 8
Collaboration and new contracts

8.1. Collaboration/project alliance/integrated project delivery

Collaboration

The *Farmer Review* (CLC, 2016, p. 33) made the comment that

> Industry-wide adoption of digitisation through media such as BIM… is predicated on collaboration. The BIM model sits at the heart of any project and only functions fully if traditional design and construction barriers are broken down by multi-party liaison and working.

Therefore, it is quite clear that collaboration is a key element of the BIM process.

Collaboration between the participants in construction projects is clearly important to the efficient delivery of built facilities. Currently, many projects are increasingly working in new collaborative environments in order to achieve higher standards of quality assurance and the reuse of existing knowledge and experience. A major constituent of these collaborative environments is the ability to communicate and reuse and share data efficiently without loss, contradiction or misinterpretation.

The main factors driving collaborative behaviour are

- trust and/or respect
- openness and communication
- common goals/new ways of working
- leadership/interpersonal skills (including an awareness of the needs of other professionals, relational skills, caring behaviour and sensitivity).

However, to be fully effective, collaboration needs to be supported by clear (to avoid misunderstanding) and accepted (to ensure they are actually followed) contractual terms. An optional or non-binding 'collaborative' understanding can give rise to a real risk that the collaborative promises will simply fall by the wayside once there is a dispute.

There are common building blocks of the requirement for collaboration in some of the current standard form contracts, including terms which place a duty on parties to act in 'mutual trust and cooperation' (NEC, 2017) or similar such terms; it must be noted that when interpreting contracts, the courts have shown a willingness to take into account such terms, even when contained in a non-binding partnering charter or similar document.

Despite the passage of time since the *Latham Report* (Latham, 1994) and the *Egan Report* (Egan, 1998) in the 1990s which called for collaboration/partnering, that ethos remains a challenge within the construction industry some 30 years later. The problem with collaboration or partnering, if you can term it a 'problem' as such, has mainly been a hesitation to adopt a partnering or collaborative approach as one or both parties feel as though they may be 'giving something away'. In the more open BIM process the parties have to accept that they will be 'giving something away' but that, in doing so, they will greatly benefit from the more collaborative approach of all participants in the BIM process.

However, the industry recognises that BIM is closely linked with collaborative working and, although in an ideal world there would be no need for a contractual obligation to collaborate, for the foreseeable future it is clear that there does need to be a contractual framework to foster such collaborative processes.

In general terms, collaboration entails finding the right group of people (skills, personalities, knowledge, work styles and chemistry), ensuring that they share a commitment to the collaborative task at hand, and providing them with the environment, tools, knowledge, training, processes and facilitation to ensure that they can work together effectively and efficiently.

In order to optimise efficiencies from a process such as BIM, a collaborative team structure ideally needs to be in place. That team structure needs to be one in which team members either are contractually obliged to (not always the ideal way) or in some other way have agreed to work in a unified manner, and will provide each other with data and information that will allow the other team members to perform their work faster, better and/or more cost-effectively.

At the same time, each team member needs to be able to insert, extract, update or modify information within the various BIM models to support and reflect their own role in the process and to ensure that each project team member remains responsible for their own element of the design and/or the data input (or not, if that is what has been agreed).

Used in this way, the BIM process serves as a catalyst for change for the relationships between the parties, and eventually for the basis of their respective agreements and appointments.

Early engagement between all prospective BIM participants is key to the successful use of the BIM process and of greater collaborative working.

The construction and infrastructure industries consist of an enormous range of professionals, specialists, subcontractors, main contractors and employers, all of varying sizes and complexity, and all as diverse as the individuals of which they are composed. They all work within a deluge of intensive information flow and under considerable time and budgetary constraints, and they are required to produce some of the most complex projects it is possible to conceive.

The BIM process offers the welcome opportunity to achieve greater cohesion, harnessing the benefits of inclusion and diversity through an improvement in the ease and speed with which those within the construction and infrastructure industries can communicate with one another and can share ideas, perceptions and interpretations of the proposed design.

Notwithstanding the diverse or fragmented (depending on your interpretation) nature of the construction and infrastructure industries, the process of design and construction always has been, and will always remain, by its very essence, a team activity. One difficulty this has presented in respect of the implementation of the BIM process is that each of the specialists' individual areas, whether professionals, contractors or subcontractors and so on, have tended to develop their own software applications in isolation to deal with the issues that they face; this obviously can go against the required collaborative mindset.

Clearly, if the BIM team is to collaborate effectively and participate successfully within a BIM environment, it is essential that the system (eventually) achieves interoperability between the various applications being used so that the product of the multiple applications may contribute effectively to the project. Ideally, and with this facility in place, the need to insert and/or dispense with information and data could be carried out by some automatic process rather than being entered manually.

To achieve interoperability generally, software providers may need to provide links or exchanges to specific software, and in doing so the same software companies will then need to provide the necessary support tailored to a certain set of specific needs. This depends largely on whether the various software providers are prepared to enter into this type of collaboration between themselves.

Importantly, design teams must recognise the benefits of openly sharing all available electronic information with the entire BIM project team. Structural analysis models, for instance, will obviously have value to other team members (including those not involved with any structural design or analysis), so delivery of these models, for example, should be part of the overall design process.

Along with the responsibility of sharing information, the designers have an obligation to convey the quality of the information that is provided and ensure that it meets the required standard. Clearly, if the geometry or the load details in a design model are not completely accurate, this needs to be made known and be well documented. In addition, the source of the incorrect information in the design documents needs to be established so that this can be traced back and rectified.

Design teams must also honestly evaluate the submittal process and work with the rest of the BIM project team to develop the best processes for the project. Together the BIM team must find appropriate ways to facilitate communication, without unduly burdening any single member with unreasonable additional liability. The request for printed shop drawings or resisting the requests of the team to distribute electronic files, simply because that is the way that the design process has traditionally been conducted, is not helpful to the project or to the growth of the use of the BIM process in the construction and infrastructure industries.

It is plain that BIM is a tool that will help the project team to communicate the needs of the project quicker and more accurately than through current practices. However, the BIM process cannot perform without the cooperation of the entire team; therefore, each BIM participant must contribute its information to the BIM process openly and for the betterment of the project as a whole.

In other words, there must be collaboration between all participants in the BIM process which is not limited because any particular participant fears that it will be 'blamed' for something that it did or did not do.

Project alliance

The project alliance approach to contracting has been successfully used for many years for oil exploration, for the delivery of infrastructure and in other fields also; therefore, there is no logical reason why it cannot be used successfully in the construction and civil engineering industries.

Project alliancing may be defined as a commercial/legal framework between a client (end-user) and one or more parties for the delivery of a project. It is characterised by

- collective sharing of project risks
- no fault, no blame and no dispute between the alliance participants (except in very limited cases of default)
- foresight applied collaboratively to mitigate problems and shrink risk
- a clear division of function and responsibility, which helps accountability and motivates people to play their part
- payment of parties providing services under a three-limb payment plan comprising: the reimbursement of the party's project costs on a 100% open-book basis and/or on a more front-loaded basis; a fee to cover corporate overheads and normal profit; and a gain/pain share process where the rewards of outstanding performance, the pain of poor performance and the financial gains and losses are shared equitably (but not necessarily equally) among all alliance participants
- unanimous principle-based decision making on all key project issues
- an integrated project team selected on the basis of the best person/organisation for each position.

Two common project alliancing features work particularly well with BIM.

- In an alliancing project, the parties agree that they will not sue each other, except for wilful default. Sharing information cannot lead to liability, and therefore the liability concerns that may impede the adoption of the BIM process do not apply in an alliancing project.
- Because a portion of compensation is tied to a successful outcome, there is an incentive for the parties to collaborate. In this context, the BIM process is an ideal platform for interactively sharing information, ideas and solutions in respect of both the design and the construction stages.

There are presently several collaborative contract approaches that are used, some of which are summarised below.

Integrated project delivery

Collaborative project approaches are beginning to appear in the USA under the general name of 'integrated project delivery'. Unlike alliancing, which has generally been used for major civil and industrial infrastructure, in the USA, integrated project delivery has been primarily used for complex structures such as hospitals.

Integrated project delivery is a radical departure from traditional prescriptive and adversarial contract approaches and offers the potential for increased project value and greater reward for all participants to manage uncertainty and risk, thus eliminating the fear that causes participants to focus on their narrow self-interests.

Under the integrated project delivery approach, a special-purpose entity is created which contracts with the architect and the contractor to design and construct the project subject to its direction, with incentives and risks based on project performance. This approach is flexible, but complex, and, depending on where the project is located, can create corporate governance and licensing issues.

An alternative approach to this is more traditional, where the contractor and the architect separately contract with the appointing party (the end-user) under terms that encourage collaboration but still adhere to traditional compensation models, risk allocation and project management roles.

Interlocking risk allocations

Interlocking risk allocations leave contract boundaries in place but lessen their importance. Under this approach, the key participants jointly negotiate specific limitations to their individual liabilities, using releases, indemnifications and limitations of liability clauses. The use of interlocking risk allocation lessens the fears of liability that can hamper the free flow of information. However, interlocking risk allocation has three potential drawbacks.

- There is a risk that the provisions will be inadequately drafted or that they will not be complete.
- Some jurisdictions have restrictions on liability limitations or indemnification that could undermine this approach.
- Although the lessening of the risk allocations may disincentivise any aversion to the collaboration required, it does not create any additional incentive to actually collaborate.

Therefore, to enhance their chances of success, interlocking risk allocations should be balanced by performance incentives.

Single-purpose entities

A single-purpose entity is a limited liability enterprise (e.g. corporation, limited liability company, limited liability partnership) created to design, construct and possibly own and operate a facility. Such entities solve the boundary problem by bringing all parties within one boundary.

The key participants sponsor the single-purpose entity and achieve a gain by optimising the entity's success. The single-purpose entity contracts with the owner or employer for the services required to construct the built asset, with the specifics of scope, responsibility and liability determined on a project-specific basis. To be fully effective, the parties within the single-purpose entity must release each other from most potential liabilities, or at least agree that any 'in boundary' claims will be paid only by project insurance.

Single-purpose entities are common in off-balance-sheet asset financed projects (project finance). Under a classic project finance structure, non-recourse loans are used to design and construct a revenue-generating asset that is owned by the single-purpose entity. The asset, and any guaranteed income streams, secures the loans. As might be expected, there are numerous variations with limited recourse and with limited sponsor guarantees and similar features. However, the fundamental economic principle of the single-purpose entity is that the owner's return is based on creating value in the single-purpose entity.

8.2. New contracts/smart contracts/blockchain

It is likely that the increasing use of the BIM process will bring with it the development of new types of contract particularly for use on BIM projects; some of the possibilities in this regard are outlined below.

As the use of the BIM process develops further to the level of fully integrated BIM, it may be that traditional bi-partite contracts, even with BIM addendums or information protocols (BIM protocols), may become unsuitable and collaborative multi-party contracts may potentially become more appropriate.

The BIM process has the potential to substantially alter the relationships between parties and blend their roles and responsibilities. With this position in mind, risks will need to be allocated rationally, according to the benefits a party will be receiving from BIM, the ability of the party to control the risks, and the ability of the parties to absorb risks through insurance or some other means.

There will be a need for standard contract documents suitable for BIM; without these the development of BIM is likely to be hindered. The emergence of BIM as a vehicle for dramatic change in design and construction occurs in a legal environment that, as yet, has not fully come to grips with all the risk management implications of the underlying technology of electronic representation, sharing and transmission of documents of any type.

Appropriate contract language that will foster the open sharing of BIM information must be developed. However, the contract language cannot alter the relationships of the project team members or change their responsibilities beyond the extent of what they are permitted to perform. Therefore, there is a clear tension between the need to tightly define responsibilities and limit reliance on others, and the need to promote collaboration and encourage reliance on information embedded in BIM models.

The terms of appointment for all members of the project team will need to be considered and drafted to reflect the collaborative nature of BIM and to ensure that their responsibilities, duties and services are aligned.

Communication and methods of working will need to be outlined to enable those who are not in direct contractual relationships to work together in order to deliver the project in the integrated and collaborative spirit required by BIM. Provisions on the input of data, limitation of liability and defining the role of the BIM information manager will also be crucial.

In addition to all of the above, any BIM contract or appointment (whether bi-partite or multi-party) will also have to define the status of the BIM model and deal with post-handover matters such as lifecycle management, data capture and the 'golden thread' of information.

One major issue that will need to be addressed is: does the party that is managing the model assume additional liabilities and risk? The answer to that question should be 'no'.

Current BIM practice uses a series of interlocking models to communicate the design and construction intent for a project. In many instances, the complete design is only visualised when imported into a viewing programme. Moreover, most models do not contain all the construction details required for a project. Thus, the contract documents will include some 2D information that is added to the information in the BIM model. In practice, these issues are currently resolved by using a printed submission as the contract document, even if the communication flow has been digital. This is not an ideal solution going forward.

There are several options currently being followed. The first is that a 'co-contract document' is used between the parties but is not submitted to permitting agencies. In this case, the appointments need to state how inconsistencies will be handled. Another option is to use the BIM model as an 'inferential document'. Under this option, the model provides visualisation of the design intent inferable from the contract documents, but this does not have any direct contractual effect. Finally, the BIM model can be used as an 'accommodation document' that can be used, but not relied on, by the recipients. This last approach is similar to the computer-aided design (CAD) transfer liability waivers that designers use when providing CAD documents/drawings to contractors. However, it must be remembered that limiting reliance between the participants in this way is likely to undermine the BIM model's full purpose and utility.

Multi-party contracts?

The potential of this option is illustrated by the projects delivered under PPC2000 (ACA, 2013) and other comparable multi-party contracts that are being developed and used in Australia and the USA. PPC2000 has a BIM addendum clarifying that

> each Party shall be responsible for any contribution it makes to a model or that arises from the party's access to that model (ACA, 2013).

Smart contracts

Although smart contracts are still being developed and are only emerging in use in some areas of industry, it is a fast-growing technology that has the potential to be used in many more situations.

The Law Commission (a statutory independent body created by the Law Commissions Act 1965 to keep the law of England and Wales under review and to recommend reform where it is needed) has defined smart contracts as

> a legally binding contract in which some or all of the contractual obligations are recorded in or performed automatically by a computer program (Law Commission, 2021).

This definition was provided by the Law Commission in its paper on smart contracts published in November 2021, itself the culmination of research into whether the UK's common law system is well-equipped to deal with smart contracts (more of which below).

A smart contract which is normally written in machine-executable code allows a given programme to execute obligations within a contract once the conditions of that contract have been fulfilled (i.e. in simple terms, if 'X' happens then do 'Y'). Therefore, when certain conditions are met, the smart contract automatically self-executes, and this process should facilitate trust, transparency and efficiency.

Automated contract performance is not a new thing, at least conceptually. Plenty of contracts are conditional on and/or are expressed automatically to come into force on certain outcomes or the performance of certain tasks. However, automated contract performance wholly or partially based on code (i.e. a smart contract) takes this to a new level (in that it is a programme that will actually and automatically be doing something as compared to a piece of paper telling the parties that something will happen when this is activated). This new approach raises interesting questions around efficiency, transparency and reliability.

Where a smart contract is to be used, the parties to that contract still negotiate the terms as usual, and those agreed terms are then transposed into a code, which can be (but is not always) accompanied by natural language clauses. Therefore, it is perfectly possible to have a smart contract acting as an execution mechanism for a particular part of a wider natural language contract (but this may not always be the case).

Due to their inherent transparency, smart contracts can reduce the reliance on trust (or mistrust) between parties. Generally, the self-executing obligations in a smart contract are stored on distributed ledgers (such as a blockchain – dealt with further below) which are designed to be difficult to alter without the consent of all of the contracting parties. Therefore, once agreed and in force, the smart contract is designed to continue to operate without further intervention.

Common applications for smart contracts presently include supply chain management, banking transactions, customer verification, peer-to-peer transacting and other repetitive tasks, although there is increasing potential for them to be used in many more complex agreements.

Despite the positive outlook for the future of smart contracts, the issue of smart contracts does raise some challenging questions for the legal profession, including, for example, whether they would be expected to be able to write the required code, how can it be identified that a smart contract is appropriate, whether the legal professional needs to understand how the smart contract fits together (or just rely on an automated process), and whether the code used is itself binding and cannot be changed.

The Law Commission's view is that the UK's common law system is well-equipped to deal with smart contracts. However, this view is accompanied by recommendations on how best to ensure that a smart contract forms a binding contract from a traditional contract law perspective, including that parties: (*a*) distinguish which definitions prevail in the event of a conflict, if expressed in both natural language and code; (*b*) provide a jurisdiction and choice of law clause; (*c*) allocate the extent of liability in the event of bugs or coding errors and (*d*) provide an explanation in natural language to accompany coding, while making it clear that the explanation forms (or does not form) part of the contract (as appropriate).

An interesting point arises also with regard to the interpretation of coded terms. The courts have developed principles of interpretation in relation to natural language, but it is unclear whether a similar interpretation principle will easily apply to terms in code. This is an interesting area of untested law.

Because of this, the Law Commission has already suggested a test whereby an expert coder could assist the court in ascertaining the meanings of coded terms – that is, a 'reasonable coder' test.

Additional uncertainties and potential difficulties will no doubt arise as the use of smart contracts continues to increase, particularly in relation to conflict of terms and conflict of law issues.

Smart contracts can be a valuable tool when applied to the BIM process, particularly (but not only) in respect of the construction and project management elements of the process.

Some of the ways in which smart contracts can be applied to the BIM process are discussed below.

Payment automation

Smart contracts can automate payments based on predefined milestones or deliverables in a construction project. When a specific phase of the BIM model is completed or validated, the contract can release payments to the responsible party automatically.

Change orders

Smart contracts can streamline the process of issuing and managing change orders by automatically updating project specifications and budgets when changes are approved, thereby reducing disputes and delays.

Asset handover and acceptance

When a building project is completed, smart contracts can automate the handover process. Once all criteria for acceptance (e.g. inspections, testing etc.) are met, the contract itself can trigger the transfer of ownership and relevant documents.

BIM model version control

Smart contracts can manage version control of various BIM models, ensuring that all project stakeholders work with the latest and approved BIM model versions. When an authorised update is made, the contract can validate it and distribute the updated model to all relevant parties.

Quality assurance and inspections

Smart contracts can automate quality assurance processes by scheduling and tracking inspections based on predefined criteria. When inspections are successfully completed, the contract can record the results and trigger subsequent actions or approvals.

Supply chain management

In BIM-related supply chain processes, smart contracts can help manage orders, deliveries and payments to suppliers and subcontractors automatically. For example, when materials are delivered and verified, the contract can automatically initiate payment.

Data security and access control

Smart contracts can enforce data security and access control in BIM projects. In this way, only authorised parties with the correct permissions can access and modify specific parts of the various BIM models or the other project data.

Dispute resolution

In case of disputes or disagreements during a project, smart contracts can automatically facilitate dispute resolution mechanisms. They can trigger mediation or arbitration processes based on predefined terms, after predetermined triggers are met, helping to resolve issues more promptly, efficiently and cost-effectively.

Maintenance and facilities management

Beyond the construction phase, smart contracts can be used for ongoing facilities management and can automate maintenance schedules, trigger work orders and manage service contracts based on predefined conditions and data from the BIM asset model.

Environmental and sustainability monitoring

Smart contracts can automate the monitoring of environmental and sustainability goals in construction projects. When predefined sustainability requirements are met, the contract can automatically trigger certifications or awards.

It is important to note that the implementation of smart contracts in the context of the BIM process requires technical expertise in blockchain (dealt with further below) and smart contract development. Additionally, project stakeholders/participants must clearly define contract terms, conditions and required indicators to ensure that smart contracts fully operate as intended. When correctly established, the use of smart contracts in the BIM process has the potential to enhance transparency, reduce disputes and improve project efficiency.

Types of smart contracts

There are already many smart contracts in use; some of the smart contracts presently in use are

- decentralised autonomous organisations (DAO): a DAO is a smart contract which is bound to specific rules that are coded into blockchain contracts combined with governance mechanisms
- application logic contracts (ALC): an ALC is another type of smart contract in blockchain which allows devices to function securely and autonomously. These contracts contain an application-based code, which typically remains in line with other blockchain contracts, enabling communication across different devices so that the Internet of Things (IOT) can merge with blockchain technology.

Smart contracts and blockchain

Set out below are some features of smart contracts and blockchain.

Security and high reliability

Transactions can be performed with high reliability. In addition, as the distributed ledger is highly encrypted, it is virtually impenetrable and offers high security.

Disintermediation

Smart contracts eliminate reliance on third-party intermediaries to perform transactions. They enable parties to enter into agreements without any dependence on intermediaries. Disintermediation creates a cohesive environment because both the logic and information in the contract are visible to all those participating in the blockchain network. Also, parties involved can check out any edits made by a party to the contract's content as it is visible to everyone involved, thereby improving transparency.

Fewer errors

As the contracts use software code to automate tasks, they are less prone to manual errors. Further, they offer real-time updates with both speed and precision.

Less human intervention

The transactions do not need much (or in some cases any) human management, which reduces the risks involved in the contract execution cycle.

Speed and efficiency

As smart contracts execute immediately when a predefined condition is met, they offer speed and efficiency. In addition, smart contracts are digital and automated, and there is no paperwork to process and no time spent reconciling errors that often result from the manually filling in of documents.

Savings

As smart contracts remove the need for intermediaries to handle transactions, this reduces their associated time delays and additional fees.

When supply chain participants use smart contracts (particularly through blockchain), they can help to streamline the flow of goods by automatically triggering the next steps when conditions are met, including shipping, delivery and other follow-on steps. Thus, participants can act earlier to eliminate disruptions should an unexpected event occur.

Smart contracts are essential to trade finance, creating an ecosystem of trust for global trade. Standard rules and straightforward trading options encoded in smart contracts decrease risk and help participants find new opportunities.

Smart contracts have been around since about 1994, when Nick Szabo, a computer scientist, developed them with digital codes in mind. Smart contracts in blockchain are digital contracts that are stored on a blockchain. They typically run after meeting certain pre-established conditions. Further, they are used to automate the execution of an agreement. So, all participants can immediately look at the outcome without third-party involvement and without any time loss. In addition, workflows can be automated, triggering the next action when the required conditions are met.

A blockchain is a decentralised digital ledger that securely records transaction data across a network. A blockchain consists of digital blocks, each containing transaction information (similar to a cell in a spreadsheet). Once a digital block is full, the data is run through an encryption algorithm, creating a unique number (referred to as a 'hash'). That hash links the current digital block to the previous digital block, forming a chain (hence 'blockchain').

Unlike traditional databases, blockchain is decentralised; therefore, no single entity controls it; instead, all users collectively maintain control. Importantly, data entered into a blockchain is irreversible. For example, in Bitcoin (a form of blockchain), transactions are permanently recorded and viewable by anyone. Blockchain ensures data integrity through cryptography (i.e. code writing) and consensus mechanisms (i.e. the complete set of ideas, protocols and incentives that enable a distributed set of nodes to agree on the state of a blockchain). Once information in a blockchain is recorded, it cannot be altered retrospectively.

Collaboration through BIM is a profound change that creates great opportunities, as noted above, but also creates new legal and liability issues. All significant construction projects in the UK incorporate a written contract at their core. At their most basic level, contracts legally record that which has been agreed on by the parties to the contract. But not all agreements are legally binding. To constitute the 'classic' contract under English law, there are essential elements that must exist (e.g. offer, acceptance, consideration etc.) and elements that must not exist (e.g. an agreement to agree etc.). Beyond those basic requirements, construction contracts can be and often are as complex as the matters within them, and are also often beyond the original anticipation and contemplation of the parties to the contract.

In reality, it is what construction contracts oblige participants to do, refrain from doing, either expressly or implicitly, or those matters on which the contracts remain silent, that have formed the basis of construction contracts being seen as adversarial.

Contracts determine the parties' risks and rewards, but also the rights and obligations of the parties, and also sometimes of the wider participants. Thus a construction contract creates a legal framework on which the parties' duties and obligations are defined.

There are various standard form contracts available (e.g. the Joint Contracts Tribunal (JCT); the New Engineering Contract (NEC)) and, while some contracts are written ad hoc for a specific project, the majority simply use one of the standard form contracts with bespoke provisions and/or alterations to the existing wording to suit the particular requirements of a project and its stakeholders.

It is often considered that the standard forms of contract (particularly those with ad hoc alterations and amendments) pit parties against one another, rather than effectively managing conflict and promoting the pursuit of amicable solutions when situations arise.

With the above in mind, because of the 'enforced' collaborative nature required by the BIM process, the construction industry has perhaps its best ever chance of ridding itself of contracts that breed confrontation, conflict and dispute through the use of smart contracts (or some other similar more advanced approach).

Traditionally, most contracts are bi-partite agreements (i.e. an agreement between two parties only) with clearly defined boundaries in terms of scope and liability. These contracts tend to insulate and isolate rather than collaborate. One contract tends to define one pair of relationships or one set of requirements in a project, each having little regard for any other.

The common law legal system is essentially individualistic, focusing on individual rights and responsibilities. Great efforts are expended on determining where the responsibility of one party ends and the responsibility of another begins. Many of the most fiercely fought battles in construction law focus on the dividing line between entities. Privity of contract, the economic loss doctrine, means and methods, and third-party reliance are all issues where drawing lines between parties is essential to determining responsibility and liability.

However, the need to consider appropriate amendments to the presently published standard forms of contract becomes more and more important when the level of BIM being used becomes more intense and, as matters progress further, fundamental changes to contracts/appointment documents will almost certainly be necessary, as the traditional legal position and relationship between the parties is likely to change in very significant ways.

8.3. Summary

- Collaboration is a key element of BIM.
- Collaborative approaches that have been used include project alliance, integrated project delivery, interlocking risk allocations and single-purpose entities.
- New contracts, smart contracts and contracts in blockchain are being developed.
- A key question will be whether the UK's common law system is equipped to deal with the proposed new contracts and smart contracts.

emerald PUBLISHING ice Publishing

Peter Barnes
ISBN 978-1-83608-515-7
https://doi.org/10.1108/978-1-83608-512-620251010
Published under exclusive licence by Emerald Publishing Limited

Chapter 9
Design, professional indemnity insurance, intellectual property rights

9.1. Design and design liability

Traditionally, the design for a built asset was based on drawings and specifications produced by an architect or another designer. The design may have included separate and individual inputs from consultants (e.g. structural engineers, mechanical and electrical engineers) and sometimes from specialist contractors and subcontractors. However, when operating under the BIM process, the design process will need to be a much more collective and collaborative approach.

Given the reduction in design barriers through the use of the BIM process, certain design elements, such as specific object properties, may in some cases be created by suppliers, specialists or even software manufacturers, rather than by what may more traditionally be considered as design professionals.

In addition, the design itself may be self-modifying, particularly if a parametric model approach applies, and to that extent it may be partially self-designing in any event.

Under the BIM process, the design deliverables required are more likely to be a computer model or simulation rather than paper drawings, and the resulting model may be distributed between computer systems operated by different participants.

Participants may also themselves add to a design model, and the design will therefore be fluid and flexible.

In addition, the complete design (or partially complete design) may exist on a cloud server rather than as a set of hard copy drawings.

The above changed concepts in respect of design bring with them certain challenges, as outlined below.

- At tender stage, the design needs to be clearly expressed and defined, as contractors need to know that they are all bidding against the same basic design information.
- The revised design elements from earlier models need to be easily identified so that variations to the original scope of works can be demonstrated.
- Inspectors and other building officials must be able to compare the physical construction with an objective and stable design standard, not with a model that constitutes a 'moving target'.

- Designers need assurance that their services are complete and, if problems occur later, that their designs can be compared against the constructed building.
- Appointing parties need to determine whether they have received a project that complies with the design.

Clearly, the fluidity in design that is allowed for by the BIM process is likely to compete with the precision required for contract enforcement. Therefore, the contract definition of design needs to address two key issues.

- The contract documents between the various parties should define the design deliverables in terms of content, time and type of electronic media to be used.
- The contract documents between the various parties should determine whether incorporated submittals, such as parametric objects provided (perhaps) by suppliers, specialists and so on are part of the designer's BIM deliverables, and which party is responsible for the incorporation and coordination of those submittals.

Once a defined design has been adopted in a contract or an appointment, it will obviously be important for the parties, and particularly for the designer(s), that this defined design is adhered to during the project development.

For the ease of future reference and to maintain a retrospective record, the design should be preserved as 'snapshots' at major design milestones (data drops). In some cases this may be accomplished by printing and saving the milestone documents. However, in a multi-dimensional electronic design maintained on a cloud server, the total design package may never be encompassed within printed documents; in such cases it may be possible to temporarily 'freeze' the digital design at a particular time and save it complete with linked documents and locations on some form of semi-permanent media.

In the meantime, and in parallel, the model containing the design will continue to evolve and will be developed by the various approved participants.

As outlined above, when the BIM process is applied, not only is the concept of 'the design' less clear, the identity of 'the designer' is equally vague. It may be that there will still be a designer who is responsible for the overall design concept and who will maintain responsibility for systems design, the overall layout of design elements, the flow through the structure and the 'aesthetic' elements of the building. However, there will also probably be many other 'designers' (including consultants, suppliers, specialist contractors and subcontractors) who will add to and (in some ways) amend the design. All these various designers will now be obliged to work together in a collaborative way.

In such a collaborative and fluid design setting, where the design input can easily be 'hidden' in the design model process, several significant questions are relevant.

- In the case of design failures, how will the various designers' contributions be extricated from the model to determine responsibility for the design failure?
- Will parties accessing the shared model be legally able to rely on the contributions of other designers, or will the legal doctrine of privity of contract (i.e. where only

the parties to a contract owe duties to one another and realise any benefits under the contract) become an issue?
- What are the responsibilities of any secondary designers?
- To what extent can the design professional rely on the products of any secondary designers?
- If the software can communicate between objects and cause them to adjust their properties, does the software itself become a 'designer' as well?
- Do the standards committees that develop interoperability protocols and object specifications also become project 'designers'?
- If these other parties and entities do have design responsibility, are they required to have insurance for the design risks?

In the immediate future, appointing parties, owners and building users and developers will almost certainly look to the normal pool of architects and engineers to be the project's designers. But, in a practical sense, these parties cannot check and be responsible for the work of the many other individual designers distributed throughout a collaborative (and possibly convoluted) design process.

In reality, most disputes regarding design deficiencies have little to do with the main design elements, but rather arise from deficiencies, discrepancies and divergences in details, inadequate coordination, deviations in submittals, excessive changes and failure to meet budgetary or functional programme requirements.

Therefore, in developing contract documents, careful thought should be given to integrating appropriate limitations (where applicable) of liability and waivers.

The normal professional registration requirements in many jurisdictions insist that a licensed professional is to be in 'responsible charge' of all work performed by its design firm, and this work must either be performed by or be supervised by the responsible professional. In fact, the appointment documents are normally sealed by the responsible professional to signify compliance with this requirement and acceptance of this responsibility.

However, if design responsibility is distributed between many different parties, is this still possible? How can a professional supervise design contributions by firms that are not under his or her control and/or how can a design professional supervise changes to structural detailing that are performed by the software itself?

In the short term, appointing parties are likely to accept sealed drawings without considering what portion of the content has actually been created under the responsible charge of the responsible signing professional. However, in the longer run, the professional registration requirements (and the applicable professional indemnity insurance cover) will need to be considered and may need to be modified to reflect the actual practices, and realities, of digital BIM design.

Even in the traditional design environment, design delegation creates potential issues with licensing and with responsibilities, and this is magnified when BIM design is used.

For example, BIM designs, especially when based on the design of specific objects that have encompassed within them their own design parameters, can contain embedded information provided by manufacturers and subcontractors.

In addition, some BIM software can react to changes in the model. Structural design software, for example, can parametrically change details in response to changes in the design. In neither of these cases will the architect or the engineer have created the information, and probably will not even have been able to check the information before it is incorporated in the BIM model.

Naturally, architecture and engineering practice will continue to evolve and use increasingly powerful design tools (some based on artificial intelligence), but the legal and regulatory structures have not yet fully adjusted to this change in practice.

There are many design interfaces between different design team members, whether between different consultants, between consultants and a contractor, or between a contractor and its own designer and/or its subcontractors and/or suppliers, and it is the lack of clarity in respect of these interfaces that is most likely to give rise to project problems and disputes.

The design processes described in professional appointments, building contracts, design programmes and project execution plans do not always clearly identify the points at which a consultant designer – whether an architect, structural engineer or mechanical and electrical consultant or otherwise – expects additional detail to be added by the contractor through contributions from its appointed subcontractors and suppliers. This situation is complicated further where the appointment of the contractor (and therefore of its subcontractors and suppliers) is not completed until the project is about to start on site.

These liability issues highlight concerns that arise from the distribution and delegation of design. Although design delegation issues can exist with or without collaborative use of BIM, they are clearly much more significant when more parties are involved and are involved more closely with each other.

As noted above, in a collaborative BIM model many parties contribute to the design. Crucial details embedded in the design may be provided not by design professionals but by specialist subcontractors or consultants. In addition, BIM software is designed to react to changes in the model by modifying elements of the design affected by a change. These circumstances increase the potential liability exposure of design professionals who use BIM collaboratively, and they risk assuming overall model responsibly.

This can lead to concerns about whether or not the use of BIM may alter the traditional allocation of responsibilities between the appointing party (client), contractors, designers and suppliers. If each party is responsible for its own model (or its contribution to that model), to what extent (if at all) is the BIM information manager/BIM coordinator liable when clashes are not detected or the design is not coordinated?

As noted earlier in this book, the role of the BIM information manager is not meant to be equivalent to a lead designer. The BIM information manager is responsible for the management of information, information processes and compliance with agreed procedures, not the coordination of design or the design itself. However, this does need to be spelt out in the information protocol or in the appointment documents, otherwise there is potential for conflict with regard to the design and the design coordination roles.

At fairly basic levels of BIM, the BIM process should not alter the traditional responsibilities between the various participants to any great degree because at that level BIM is largely a series of federated models prepared by different design teams (the number of models and their purpose being determined by the appointing party) and put together in the context of a common framework for the purpose of being used for a single project, with licences granted to other project team members to use the information contained in the federated models. Therefore, if each model is thought of as a drawing or design in the more traditional sense, then, provided that the individual's appointment clearly defines the role and responsibilities of the designer in the usual way, there should not be any significant change.

Although the usual responsibilities will remain in the above situation, it will be important that the design brief is understood and that the design is reviewed as an ongoing obligation. The new BIM technology and the new way of producing design should not, in and of itself, change the fundamental legal principles. Consequently, there will still be the need for a BIM coordinator to coordinate the design, a role that would previously have normally been undertaken by the lead designer.

The various appointments should be very clear that a design consultant has no authority unilaterally to delegate design responsibilities. Therefore, clear agreement on the matrix of design contributions and their timing through the use of BIM models should be a more effective means of defining and limiting a consultant's potential liability.

The issue of delegated design responsibility has been subject to some case law guidance over the years on how the overall cost of damages claimed as a result of a failure of design should be shared between the respective designers, and the use of 'net contribution' clauses in this regard is commonplace. A net contribution clause follows the general principle that the contributions should be 'just and equitable' with regard to the extent of that person's responsibility for the damage caused; this general principle is likely to remain the case in respect of the BIM process for the foreseeable future.

Clearly, the use of BIM will not resolve the above issues which will largely remain in place and, in some instances, the BIM process may add to these problems; however, if the BIM process reduces drawing errors and mis-communication between the parties in any significant way, which it is intended to do, then it is likely that the frequency and severity of loss will be lessened, even if the number of potential reliant and liable parties is significantly expanded.

9.2. Intellectual property rights and copyright

Intellectual property (IP) refers to creations of the mind, such as inventions; literary and artistic works; designs; and symbols, names and images used in commerce. Intellectual property is protected in law by, for example, patents, copyright and trademarks, which enable people to earn recognition or financial benefit from what they invent or create.

When individuals create material that forms part of a BIM model or a system they will often generate intellectual property.

The appointing party generally requires use of this intellectual property but companies working on a project and contributing to the intellectual property will naturally have an interest in retaining a reasonable degree of control over it.

When material that has been created by third parties is necessary to a BIM model or system then organisations will need to ensure that they have the intellectual property rights to use that material, making intellectual property rights a vital element in the BIM process.

The intellectual property rights created through the BIM process go beyond copyright of design by creating rights in software code, databases and the models themselves. The intellectual property provisions will need to be wide enough to allow for these alternatives and will also need to include the transmission of electronic information and to allow the material to be used in a model connected to the project. Provisions should also be inserted in appointments/contracts that deal with ownership of the various BIM models and, in particular, the final federated BIM model.

To avoid liability for infringement of third-party intellectual property rights, the information protocol (previously known as the BIM protocol) or the appointment/contract documents themselves should ensure that all contributors warrant that they hold the intellectual property rights over the contributions they make to the various models and that they provide an indemnity to all other parties who may use such contribution in the event of a third-party intellectual property dispute.

Generally, intellectual property law in the UK recognises intellectual property rights (IP rights) for each distinctive contribution and, as noted above, this should be preserved in the information protocol and/or the various appointments/contracts, as applicable.

Therefore, the information protocol should ensure that each contributor provides a non-exclusive licence and sub-licence where required to cover all members of the BIM project team, allowing access and use of each contribution of design for the purposes of the project.

These licences would be subject to the right to suspend or terminate, particularly in the situation where non-payment suspension or termination rights arise under the information protocol or the appointment/contract documents.

Taking all the above points into consideration, when the BIM process is being applied, enhanced attention needs to be given to the intellectual property provisions of the information protocol and/or the appointment/contract documents. This is especially important because the BIM model will not only show the results of patented processes and designs but will (in all likelihood) actually 'know' the building codes, algorithms and applicable engineering principles. It is this base information which the system will apply to enable the model of the building to be manipulated and updated. Therefore, in this context, design participants will be required to impart much more intellectual property on a BIM project than they are traditionally used to.

A question that needs to be considered is: as an 'intelligent model' and/or a 'parametric model' is an inherently collaborative work, to what extent can anyone claim ownership of the intellectual property in respect of those models in any event?

Because the appointing party will normally have access to the models as they are being developed, care must be taken to ensure that the intellectual property rights are not lost because of the open and collaborative nature of model development.

BIM models may also contain confidential or trade secret information. For example, a model for a manufacturing plant may disclose what a company is planning to build and the processes it will use. If such information is broadly circulated in a collaborative team, the issue to be addressed is how will this information be protected legally and practically?

Of course, many of the intellectual property issues arising during the BIM process are similar to those that existed before BIM was used; however, those issues are amplified by the increased amount of information contained in the BIM process, the requirement for more access to that information by others and its ease of transfer.

Confidentiality issues can arise subtly when the embedded information itself is confidential, although the overall design is not inherently confidential. The upshot is that who owns the BIM model, who owns information in the BIM model, and who has access to the BIM model should be considered when BIM procedures are developed and when intellectual property rights and confidentiality issues are considered.

The precise terms of any assignment or licensing of intellectual property rights will be of great concern to all design participants (particularly those in 'niche' design areas). Increased integration and collaboration also mean that there will be complex layers of intellectual property, provided by different design participants, which will be difficult to identify and reverse engineer.

Copyright is an intellectual property right which aims to stop the copying of works by others. It provides copyrights holders, such as software engineers and designers, with a range of rights in respect of their works to restrict how their works are reproduced by other people.

In the UK, provided that the work is 'original', copyright will rise automatically as soon as the work is created and fixed in material form without any need for registration. However, it is important to note that ideas are not protected by copyright, only the expression of those ideas as fixed in a material form are protected. Further, moral rights are integral parts of copyright law that protect relevant parties' personal interests in their creative works.

The key moral rights, in the context of the BIM process, are the right to integrity (i.e. authors are safeguarded against alterations or distortions of their copyrighted works that harm their reputation or honour, and this right ensures that the original work remains intact) and the right to object to false attribution (i.e. authors can prevent being falsely identified as the creators of works they haven't contributed to). In the UK, these rights are enshrined in the Copyright, Designs and Patents Act 1988 and aim to balance economic interests with authors' personal rights.

Copyright provides the owner with the exclusive right to copy, issue copies, rent or lend, perform, show, play, communicate or adapt the copyright work. Copyright is infringed by anyone who carries out any of the copyright owner's exclusive rights without their permission, unless an exception to copyright applies.

Ownership and licensing of IP rights and copyright

In respect of BIM models (and other BIM documents), in most cases, the end-user or appointing party will normally want exclusive ownership of the models and the model outputs. Any limitations on the use of the designs in the BIM models can be set out in the terms of the appointment and/or the information protocol.

Licences for the designs will be needed which should provide for limited, non-exclusive licences between parties for use of their respective BIM contributions for the purposes of the project as this will facilitate the practical development of the models and the eventual construction of the project. These licences will also facilitate collaboration as the parties do not need to be concerned with impractical claims of copyright infringement; this also facilitates an open exchange of information. The appointing party (normally the client and/or end-user) will also need limited licences to enable use of the information for the ongoing maintenance, alteration or extension of the built asset. Indemnities also need to be put in place for infringement of the intellectual property rights.

9.3. The information protocol and intellectual property rights

The information protocol for use with BS EN ISO 19650-2 (BSI, 2018b) (referred to at Chapter 7) deals with the question of ownership and licensing of intellectual property rights and copyright in the following way.

If there are provisions in the appointment in relation to copyright, moral rights, rights in design, database rights or any other intellectual property rights, those provisions are maintained but varied so far as is necessary to

- apply to any drawings, calculations, specifications and other relevant documents provided by a party under or in connection with the information protocol and any proprietary work contained in or extracted from that material
- enable the said party to grant licences or sub-licences to other members of the delivery team (as appropriate).

However, if there are no provisions in the appointment in relation to copyright, moral rights, rights in design, database rights or any other intellectual property rights, then the following provisions apply.

- All rights in the material prepared and/or provided by the party in question under or in connection with the information protocol and any proprietary work contained in or extracted from that material shall, as the case may be, vest or remain vested in the party concerned.
- The party producing the material referred to above shall grant a non-exclusive royalty-free and irrevocable licence and/or sub-licences (as applicable) to allow the material to be transmitted, copied, reproduced and used for the purposes as set out in the information protocol.

9.4. Standard form contracts and intellectual property rights

Under some of the more commonly used standard form contracts, the issue of IP rights and copyright and use is dealt with as follows.

JCT contracts

- Subject to any rights in any designs, drawings and other documents supplied to the contractor for the purposes of the contract by or on behalf of the employer, all rights including (without limitation) copyright in all contractors' design documents shall remain vested in the contractor.
- Subject to all sums due and payable under the contract to the contractor having been paid, the employer shall have an irrevocable, royalty-free, non-exclusive licence to copy and use the contractor's design documents and to reproduce the designs and content of them for any purpose relating to the works.
- The licence shall be assignable to any owner from time to time of the works or any part of them and may be sub-licensed to any owner or tenant of the works or part of them.
- The contractor shall not be liable for any use by the employer of any of the contractor's design documents for any purpose other than that for which they were prepared.

NEC4 contracts

The client may use and copy the contractor's design for any purpose connected with the construction, use, alteration or demolition of the works and for other purposes as stated in the contract, unless otherwise stated in the contract. Also, the contractor is to obtain from its subcontractors equivalent rights for the client to use material provided by the contractor.

Under NEC4 option X10, the client owns the information model and the contractor's rights over project information (as defined), except as stated otherwise in the information model requirements.

9.5. Artificial intelligence and intellectual property rights

In respect of artificial intelligence (AI), it is impossible to know whether there are in existence the correct copyright permissions for the data on which AI results are based. It cannot be assumed that there are no copyright issues as some data may have restrictions on reuse or reliance, and this would need to be taken into account when an AI model is trained in order to avoid accidental copyright breaches, this being particularly so where public AI models are used.

In the USA the courts have found that, depending on the circumstances of a particular case, no copyright is attached to an AI result where there was insufficient human input – that is, copyright cannot attach to something purely produced by an AI model.

9.6. Warranties and indemnities

Parties need to consider at an early stage whether designers should provide a warranty (including a collateral warranty), indemnity or assurance as to the integrity of the electronic data in the design model or, at least, whether they should be obliged to meet a standard of care for checking the integrity of the data and for avoiding hacking or unintended amendment to the data.

The current position with standard form contract drafting bodies is that there is no difference between the position regarding warranties/indemnities in a BIM project and those in more traditional projects. However, one factor which may be worth taking into account is the role of BIM information manager, bearing in mind that the purpose of a collateral warranty is to create a direct contractual link between the parties to it.

It may be the case that the role of the BIM information manager will be delivered as part of an existing appointment. Clearly, if that is the case, and the BIM information manager role is simply an add-on to services being provided by a party who is already providing a collateral warranty, there would potentially be no need for a further warranty from that party. However, if the role develops such that the BIM information manager is appointed under a separate appointment, then it is likely that a separate collateral warranty would be required.

9.7. Insurance

If the BIM framework is used as a collaborative framework, as should be the case, it does create possible insurance issues, and an insurance product better tailored to collaborative projects should be used (if such a product is available).

Many current professional liability policies for design have exclusions for 'ordinary construction means and methods' and for joint venture liability.

It is therefore important to ensure that the respective parties take out the appropriate insurance to cover their engagement in the BIM process.

The 'ordinary construction means and methods' exclusions, as referred to above, are designed to eliminate matters relating to ordinary construction means and methods from design liability.

However, in a collaborative setting, the designers may assist in developing sequences and construction procedures that, at the very least, skirt this exclusion. Sharing risk and reward, a mainstay of integrated project delivery, is also a joint venture characteristic, and may lead insurers to deny or limit liability if joint venture liability is alleged.

Of course, contractors also face insurance issues. Most standard commercial general liability construction policies exclude professional services and do not cover pure economic loss. Clearly, as contractors become more deeply embedded in the design process, they must consider whether they should obtain contractor's professional liability coverage (which many now do because of the design and build contracts that they are engaged on). Also, contractors must recognise that their existing standard insurance policies (even their design insurance policies) provide little (or in some cases, no) protection for 'fitness for purpose'-based claims or for economic loss claims based on their negligent performance.

Even hosting data can create additional insurance issues. Essentially, data loss more closely relates to valuable paper coverage than traditional construction policies. In addition, if the parties are developing custom software for others to use, there are product risks involved that may not be covered by their usual insurance policies. Unfortunately, software is not perfect, and residual flaws will remain despite strenuous debugging efforts. Fortunately, the software bugs are more often than not simply annoying rather than harmful; however, that is not always the case, and it could be that errors in BIM software cause economic loss to a user but the injured party has no obvious remedy for that loss through an insurance policy. However, the user's liability to other parties is not normally similarly limited, and this issue may cause a liability gap if the errors cause deficiencies in plans or other deliverables.

Of course, the insurance industry is aware of these issues and others in respect of the application of the BIM process, and this may result in the insurance market providing coverage for collaborative projects (possibly with high premium rates). However, until that time, participants in the BIM process must work with their insurance brokers to ensure that the tasks they are undertaking on BIM projects are adequately covered by the policies that are available.

Few insurance companies currently offer BIM-specific products, mostly because the use of the BIM process is still not 'mainstream', while the still relative rarity of BIM projects to date means that great uncertainties remain about its benefits and risks. Consequently, insurers are likely to increasingly offer BIM insurance policies but, until it is clear what risks are involved, premiums are likely to be high. Uncertainty is also likely to impact on the bond market until the BIM process becomes more commonplace.

Naturally, any convergence of the design and construction processes signals the need for contractors to review their overall risk profiles. Therefore, it is strongly advised that any contractor looking to participate in the BIM process consults with an insurance advisor to examine any potential increase in risk as well as the appropriateness of its current insurance policies. A thorough review of the contactor's involvement or intended involvement in the design process should be undertaken to assess the potential exposures presented.

Beyond the design/professional services liability issues that may arise from participation in the BIM process or that may exist with current project delivery methodologies, the contractor may very well have exposure to other electronic data/technology issues, such as project management software, project web hosting/web sharing, transmittal of electronic viruses or intellectual property ownership/infringement, to name but a few. These potential exposures should also be discussed with legal and insurance advisors to address both the contractual and risk-financing options of these and other technology and/or intellectual property-related exposures.

In reality, the potential problems of not being able to obtain adequate insurance cover for the various participants in the BIM process may be a major factor that impedes the development of BIM, and this is an issue that needs to be urgently addressed.

Based on the above, the pragmatic advice for parties operating in a BIM environment would be to discuss the above issues with their insurance broker or insurance company since it is likely to be a material fact relevant to coverage and it is far too important a factor to simply ignore.

Surety bond industry

The utilisation of BIM technology by the construction industry will obviously be an 'evolving' underwriting process for the surety bond industry. At present, BIM technology is such a new concept to the surety bond industry that no clear industry opinion has emerged, and individual surety organisations are likely still formulating their own respective positions. This is especially true because little guidance in the form of industry standards exists concerning the proper utilisation and application of BIM technology on construction projects. Some contractors, as early participants, are utilising BIM technology within their own organisations; such internal use by contractors does not bring any additional burden to the surety as its use is not made part of the external appointment/contract documents in those circumstances.

However, where contractors utilise BIM technology as part of their external contractual performance requirements for a project, certain surety issues are almost certain to arise, particularly in the absence of a clear contractual delineation between the respective parties for design, coordination and communication responsibilities.

Utilisation of BIM technology on projects requires a clear understanding of each party's responsibilities. The parties will need to define and address the data standards and protocols, the preparation of the model and the distribution of information from the model. In this regard, a contractor and its surety company should consider how the contractor's organisation can best manage the risks associated with utilising this new technological tool.

Although BIM technology should decrease conflicts and make the design and construction process more efficient in the long run if it is clearly defined contractually, it is still too early to attempt to predict what (if any) impact the use of the BIM process will have on the surety bond requirements of major construction and civil engineering projects; for this reason, at the moment, matters are very much considered on a case-by-case basis.

Integrated project insurance (IPI)

In the integrated project insurance (IPI) process, a single insurance policy is taken out by the appointing party to cover the risks associated with delivery; this consolidates all policies held by the appointing party and all other BIM participants.

In cases where project-wide insurance is available, this should be adopted; the UK Government has proposed the greater use of integrated project insurance in respect of BIM projects.

However, where such cover is not available (which presently is often the case), parties should consider obtaining traditional insurance coverage to protect against liability for errors or omissions in their contributions to the model.

In the case of integrated project insurance, insurers will normally appoint a technical assessor and a cost assessor to help manage their risks. With the assistance of these assessors, the insurers and the appointing party will set the project brief, assess the technical aspects of the design and its 'buildability' and establish a cost plan. The assessors monitor the work throughout the project to ensure they accord with the strategic brief and cost plan. In return for this level of involvement, insurers will structure the premium based on a reduced level of risk (because their exposure is mitigated by the presence of the assessors).

Of course, in the future, the insurance market may move towards integrated project insurance where all parties are covered on a no-fault basis, with each party contributing a percentage of the premium. This is apparently available in the UK insurance market now, although on a very limited basis, and this means that the premiums for the policies are expensive and are limited in terms of the extent and level of cover available.

Where it is available, the potential benefits of integrated project insurance include

- removing the need for the project team to adopt a defensive approach in their work
- avoiding the potentially unfair results which can occur from joint and several liability
- in some cases, avoiding the need for collateral warranties
- avoiding issues arising from a project team member's insolvency.

9.8. Summary

- The concepts of design and design liability need to be considered in the context of BIM projects.
- The position of intellectual property rights needs to be considered in the context of BIM projects.
- The position of copyright needs to be considered in the context of BIM projects.
- How artificial intelligence interacts with intellectual property rights is an important matter that needs to be addressed.
- The types of warranties and indemnities required relating to the BIM process need to be given due consideration.
- How BIM projects can be properly covered by insurance is a matter that is of concern in the BIM community.

Peter Barnes
ISBN 978-1-83608-515-7
https://doi.org/10.1108/978-1-83608-512-620251011
Published under exclusive licence by Emerald Publishing Limited

Chapter 10
Information security

10.1. General

Information security threats come in many different forms. Some of the most common threats are software attacks, theft of intellectual property, theft of identity, theft of equipment, theft of data or information, sabotage and information extortion. Viruses, worms, phishing attacks and Trojan horses are a few common examples of software attacks.

The theft of intellectual property has been an extensive issue for many businesses in the information technology field. Clearly, if confidential information about an organisation's customers or finances or new product line fall into the hands of a competitor or a hacker, that organisation and potentially its customers could suffer widespread, possibly irreparable, financial loss, as well as damage to the organisation's reputation.

However, from a business perspective, information security must, of course, be balanced against cost.

Information security involves the protection of information by avoiding and/or reducing risks to the information and forms part of information risk management.

Information security typically involves preventing or reducing the probability of unauthorised or inappropriate access to data or the unlawful use, disclosure, disruption, deletion, corruption, modification, inspection, recording or devaluation of information. It also involves actions intended to reduce the adverse impacts of any such incidents.

The information to be protected may take any form, whether electronic or physical, tangible (e.g. paperwork) or intangible (e.g. knowledge).

The primary focus of information security is to balance the protection of data, taking into account confidentiality, integrity and availability (sometimes referred to as the CIA triad), while at the same time maintaining a focus on efficient policy implementation, without unduly hampering productivity of the organisations involved.

The CIA triad of confidentiality, integrity and availability is a common model that forms the basis for the development of information security systems, segmenting these three ideas into separate focal points. This differentiation is helpful because it helps guide security teams as they pinpoint the different ways in which they can address each concern.

Ideally, when all three standards have been met, the security profile of the organisation is stronger and better equipped to handle threat incidents.

Confidentiality

Confidentiality involves the efforts of an organisation to make sure data is kept secret or private. To accomplish this, access to information must be controlled to prevent the unauthorised sharing of data, whether that be intentional or accidental. A key component of maintaining confidentiality is making sure that people without proper authorisation are prevented from accessing information and data, while at the same time ensuring that those who need to have access to that information and data have the necessary privileges. To ensure these policies are followed, stringent restrictions need to be in place to limit who can see what.

There are several ways confidentiality can be compromised. This may involve direct attacks aimed at gaining access to systems the attacker does not have the rights to see. It can also involve an attacker making a direct attempt to infiltrate an application or database so they can take data or alter it.

However, not all violations of confidentiality are intentional. Human error or insufficient security controls may be to blame as well. For example, someone may fail to protect their password, users may share their credentials with someone else, or they may allow someone to see their login while they enter it.

To fight against confidentiality breaches, restricted information and data can be classified and labelled, data can be encrypted and multi-factor authentication systems can be used.

Many, if not all, contracts have clauses dealing with confidentiality. Some may restrict sharing of any project data to any individuals outside the project team; others are less restrictive and may prohibit sharing of any project data outside the contracting party organisation.

Confidentiality clauses will usually allow disclosure in specific circumstances, such as where the data is already in the public forum or required to be disclosed by a court of law.

It must be noted that inputting data into a public AI model is, in essence, disclosing this data to the public without restriction. It is not realistically possible to delete or remove it and, as the AI model uses this data to train itself through algorithms, the AI results for other parties could be based on this data or reveal it to such other parties. This could be in breach of contracts as well as revealing confidential business information to external parties.

Integrity

Integrity involves making sure that the information and data are trustworthy and free from tampering. The integrity of information and data is maintained only if the data is authentic, accurate and reliable.

Compromising integrity may be done intentionally, but integrity may also be violated by accident. If the security policies, protections and procedures are inadequate, integrity can be violated without any one person in the organisation being held accountable.

The integrity of information and data can be protected by the use of hashing (i.e. converting data into a fixed-length string of letters and numbers), encryption, digital certificates or digital signatures.

Information security systems typically incorporate controls to ensure their own integrity, in particular protecting the core functions against both deliberate and accidental threats.

Multi-purpose and multi-user computer systems aim to compartmentalise the data and processing such that no user or process can adversely impact on another. These controls may not succeed, however, and it is not uncommon to see incidents such as malware infections, hacks, data theft, fraud and privacy breaches.

More broadly, integrity is an information security principle that involves human/social processes and commercial integrity as well as data integrity. As such, it touches on aspects such as credibility, consistency, truthfulness, completeness, accuracy, timeliness and assurance.

Availability

Even if data and information are kept confidential and their integrity is maintained, this is clearly of no use unless they are available and accessible to those who need to use the data and information. This means that systems, networks and applications must function as they should and when they should. Also, users with access to specific data and information must be able to consume it when they need to and without undue delay.

For any information system to serve its purpose, the information must be available when it is needed, irrespective of any security measures in place.

This means the computing systems used to store and process the information, the security controls used to protect it and the communication channels used to access it must function correctly and must be accessible.

High availability systems aim to remain available at all times, preventing service disruptions due to power outages, hardware failures and system upgrades. Ensuring availability also involves preventing denial-of-service attacks, such as a flood of incoming messages to the target system, essentially forcing it to shut down.

In respect of information security, availability can often be viewed as one of the most important parts of a successful information security programme. Ultimately, end-users need to be able to perform job functions; by ensuring availability, an organisation is able to perform to the standards that its stakeholders expect.

Structured risk management process

Broadly speaking, risk is the likelihood that something bad will happen that causes harm to an informational asset. A vulnerability is a weakness that could be used to endanger or cause harm to an informational asset. A threat is anything (human-made or an act of nature) that has the potential to cause harm to an informational asset.

The likelihood that a threat will find a vulnerability to cause harm creates a risk, and it is therefore important to have a risk management plan in place.

The process of risk management is an ongoing, iterative process; it must be repeated continuously. The business environment is constantly changing and new threats and vulnerabilities emerge every day.

The choice of controls used to manage risks must obviously strike a balance between productivity, cost, effectiveness of the countermeasure and the value of the informational asset being protected, bearing in mind that any processes chosen have limitations, particularly as security breaches are generally relatively rare and emerge in a specific context which may not be easily or ever duplicated. Therefore, any process or countermeasure should itself be evaluated for vulnerabilities.

It is not possible to identify all risks, nor is it possible to eliminate all risks, and there will always be a residual risk remaining.

A risk assessment is carried out by a team of people who have knowledge of specific areas of the business. Membership of the team may vary over time as different parts of the business are assessed. The assessment may use a subjective qualitative analysis based on informed opinion or may use some form of quantitative analysis.

The following items should be examined during a risk assessment

- the security policy
- the organisation of information security
- asset management
- human resources security
- physical and environmental security
- communications and operations management
- access control
- information systems acquisition, development and maintenance
- information security incident management
- business continuity management
- regulatory compliance.

In general terms, the risk management process will

- identify assets and estimate their value, including people, buildings, hardware, software, data (electronic, print, other), supplies
- conduct a threat assessment, including acts of nature, acts of war, accidents, malicious acts originating from inside or outside the organisation
- conduct a vulnerability assessment and, for each vulnerability, calculate the probability that it will be exploited; evaluate policies, procedures, standards, training, physical security, quality control, technical security
- calculate the impact that each threat would have on each asset, using both a qualitative analysis and a quantitative analysis

- identify, select and implement appropriate controls; provide a proportional response, while considering productivity, cost effectiveness and value of the asset
- evaluate the effectiveness of the control measures and ensure the controls provide the required cost-effective protection without discernible loss of productivity.

For any given risk, management can choose to accept the risk based on the relative low value of the asset, the relative low frequency of occurrence and the relative low impact on the business, or may choose to mitigate the risk by selecting and implementing appropriate control measures to reduce the risk.

Alternatively

- decide how to address or treat the identified risks – that is, by avoiding, mitigating, sharing or accepting those risks
- where risk mitigation is decided on and is required, select or design appropriate security controls in respect of the information and data and implement them
- monitor the activities taken and make adjustments as necessary to address any issues, changes or improvement opportunities.

10.2. Security controls and BS EN ISO 19650-5

There are three primary roles to think about in terms of baseline security for design and construction projects.

The appointing party, the BIM information manager and the delivery team members need to understand the types of security risks in a project or organisation, familiarise themselves with the standards required, consider the approach to file naming and document control and recognise that all data and information should be protected and be prepared to work in that way.

When planning to host a CDE (or some other document management system), a BIM participant should seek security assurances, plan a clear, navigable folder structure to aid consistency and allow controls to be applied, consider access controls and permissions, their application and monitoring, and plan for the transfer of project information in a secure manner.

After being appointed, a BIM participant should be clear in any BIM execution plan response that security requirements are understood and demonstrate how they are intended to be met, and always be aware of security requirements and abide by them. This includes making sure new team members are appropriately briefed.

Selecting and implementing proper security controls will initially help an organisation bring down risk to acceptable levels.

A risk assessment should be carried out first, followed by a security control selection. Security controls can vary in nature, but fundamentally they are ways of protecting the confidentiality, integrity or availability of information.

The security controls can be considered under the following three headings.

Administrative controls

Administrative controls (also called procedural controls) consist of approved written policies, procedures, standards and guidelines. Administrative controls form the framework for running the business and managing people. They inform people on how the business is to be run and how day-to-day operations are to be conducted. Other examples of administrative controls include the corporate security policy, password policy, hiring policies and disciplinary policies. Laws and regulations created by government bodies can also be a type of administrative control because they instruct the business in what it can and cannot do in some areas. Administrative controls form the basis for the selection and implementation of logical and physical controls. Logical and physical controls are manifestations of administrative controls, which are of paramount importance.

Logical controls

Logical controls (also called technical controls) use software and data to monitor and control access to information and computer systems. Passwords, network and host-based firewalls, network intrusion detection systems, access control lists and data encryption are examples of logical controls. An important logical control that is frequently overlooked is the principle of least privilege, which requires that an individual, programme or system process should not be granted any more access privileges than are absolutely necessary to perform the task.

Physical controls

Physical controls monitor and control the workplace environment and computing facilities. They also monitor and control access to and from such facilities and include doors, locks, heating and air conditioning, smoke and fire alarms, fire suppression systems, cameras, barricades, fencing, security guards, cable locks and so on. Separating the network and workplace into functional areas is also a physical control. An important physical control that is frequently overlooked is separation of duties, which ensures that an individual cannot complete a critical task by himself or herself, but rather needs a second authorisation.

As noted earlier in this book, BS EN ISO 19650-5 (BSI, 2020b) deals with a security-minded approach to information management. It specifies the principles and requirements for security-minded information management at a stage of maturity described as 'building information modelling (BIM) according to the BS EN ISO 19650 series', and as defined in BS EN ISO 19650-1 (BSI, 2018a), as well as the security-minded management of sensitive information that is obtained, created, processed and stored as part of, or in relation to, any other initiative, project, asset, product or service. It addresses the steps required to create and cultivate an appropriate and proportionate security mindset and culture across organisations with access to sensitive information, including the need to monitor and audit compliance.

The approach outlined is applicable throughout the lifecycle of a built asset, product or service, whether planned or existing, where sensitive information is obtained, created, processed and/or stored.

BS EN ISO 19650-5 (BSI, 2020b) provides a framework to assist organisations in understanding the key vulnerability issues and the nature of the controls required to manage the resultant security risks to a level that is tolerable to the relevant parties. Its use should not in any way undermine collaboration or the benefits that BIM, other collaborative work methods and digital technologies can generate. Implementation of the measures outlined in BS EN ISO 19650-5 (BSI, 2020b) will assist in reducing the risk of the loss, misuse or modification of sensitive information. The measures can also be applied to protect against the loss, theft or disclosure of valuable commercial information and intellectual property as well as personal data.

10.3. Data protection laws

Data protection laws are those laws that are applicable to the parties and/or to the rights, responsibilities and/or obligations of any party in connection with their appointments and (particularly in the UK) include

- the General Data Protection Regulation (GDPR)
- the Data Protection Act 2018
- any other applicable law relating to the processing, privacy and/or use of personal data, as applicable to any BIM participant and/or to the rights, responsibilities and/or obligations of those participants in connection with their respective appointments, and any laws which implement any such laws
- any laws that replace, extend, re-enact, consolidate or amend any of the foregoing.

The General Data Protection Regulation (GDPR)

GDPR is a UK law (based on EU law) with mandatory rules for how organisations and companies must use personal data with integrity. GDPR imposes several obligations and responsibilities on data controllers (i.e. those responsible for determining the purposes and means of processing personal data and data processors) and data processors (i.e. those who process personal data on behalf of data controllers).

In the BIM process, these entities may include appointing parties (clients/end-users), architects, engineers, contractors or software providers. Depending on the circumstances and on their role, those participants will need to comply with GDPR requirements, such as obtaining consent from data subjects for processing their personal data; providing transparent information about how their personal data is processed, stored and shared; implementing appropriate security measures to protect personal data; respecting the rights of data subjects to access or erase their personal data; reporting any data breaches or incidents to the relevant authorities and data subjects; and keeping records of their data processing activities.

Article 5 of the (UK) GDPR sets out seven key principles

- lawfulness, fairness and transparency
- purpose limitation
- data minimisation
- accuracy
- storage limitation
- integrity and confidentiality (security)
- accountability.

These principles lie at the heart of the approach to processing personal data; any failure to comply with the above principles may leave a party liable for substantial fines. Article 83(5) of the (UK) GDPR (ICO, 2016) states that infringements of the basic principles for processing personal data are subject to the highest tier of administrative fines, and this could mean a fine of up to £17.5 million, or 4% of the party's total worldwide annual turnover, whichever is higher.

Complying with GDPR can be challenging for BIM projects and workflows which may include coordinating and communicating with multiple data controllers and multiple data processors across many different countries and jurisdictions, while also ensuring that they adhere to the same standards and obligations. In addition, of course, there is a need to balance collaboration and integration with security and control, as well as finding the right tools and platforms that support both.

Achieving GDPR compliance for the BIM process is not a one-time event, but an ongoing process that requires planning, monitoring and reviewing.

10.4. Summary

- Information security threats come in many different forms.
- The primary focus of information security is to balance protection of data while taking into account the CIA triad (confidentiality, integrity and availability).
- A structured risk management process needs to be put in place.
- A risk assessment needs to be carried out.
- Security controls under the BS EN ISO 19650-5 (BSI, 2020b) standard need to be complied with in line with good practice.
- Due consideration must be given to the relevant data protection laws, and in the UK in particular, to the General Data Protection Regulation (GDPR).

Peter Barnes
ISBN 978-1-83608-515-7
https://doi.org/10.1108/978-1-83608-512-620251012
Published under exclusive licence by Emerald Publishing Limited

Chapter 11
Dispute resolution

11.1. General

Disputes in the construction industry can have a major impact on organisations and projects.

The impact of construction disputes can be summarised as follows

- time delays and cost overruns
- additional expense in managerial and administrative costs and time
- rework and relocation costs for operatives, equipment and materials
- loss of profitability and maybe even business viability
- possibility of court or tribunal cases
- possible loss of company reputation
- possible loss of professional reputation
- reduced respect between parties, deterioration of the relationship and a breakdown in cooperation.

Some factors that give rise to disputes in the construction process are

- failure to plan and schedule adequately
- failure to follow a plan and schedule
- failure to supply adequate resources
- disagreement over what material is specified
- an inability to agree on what constitutes a variation and what the value of the variations should be.

Traditionally, construction project records are poorly maintained, making it difficult to prove the cause and effect of a change on a project, especially during a retrospective analysis.

However, within a BIM environment, a certain level of information should be generated and stored at different stages of the project to enable successful management. Software packages support this process by enabling contemporaneous digital record keeping, such as when updating data on tablets on site.

In view of the above, the implementation of the BIM process presents an opportunity to nip in the bud any disputes before they have the opportunity to develop, and also provides contemporaneous information that should greatly assist in the resolution of a great many other disputes.

The BIM process provides the ability to view the programme implications of proposed alterations both retrospectively and, perhaps more importantly, as a projection of future actions. If parties have early warning of the potential programme implications of proposed changes, the probability of a dispute arising will be greatly reduced.

If a dispute regarding programming and/or time matters does arise, it should be easier for a tribunal to resolve such a dispute by referring back to the frozen snapshot views of the programme dimension of the BIM model. While there may still be disagreements about the reason for the delays suffered, at least there should be less dispute about the progress on the project at any particular time. One of the major difficulties in any dispute resolution process is establishing what actually happened and when, and it is plain that the BIM model should greatly assist in this particular issue.

If project information is stored electronically in a common data environment and in a structured way, the process of finding information becomes much easier because it can be searched, sorted and filtered. Critically, if project information is embedded in, or linked to, a virtual model, it can be used as a central reference point to locate all relevant information, creating a clear audit trail, which is particularly useful when trying to find information related to a certain aspect of work.

The required specification for the materials to be used will be embedded within the BIM model and there can therefore be no dispute at a later date about what that specification is. This is in contrast to the present (non-BIM) situation, where there are often disputes as to whether the material specifications on a drawing, on a schedule or on a data sheet (where these are different from one another), apply.

The BIM model contains information, or can link to information, necessary to generate bills of materials, size and area estimates, productivity, materials cost and related estimating information. It avoids the need to process material take-offs manually and thus reduces error and misunderstanding. Moreover, the linked cost, productivity and resource information evolves in step with the design changes made, and those changes can be recorded as snapshots of the BIM model if required. Therefore, if there is a dispute at a later stage, as BIM technology develops further it is conceivable that the BIM model can be unwound to the point where a problem begins to occur, and the source of the problem can then be more easily identified. Also, BIM, through its capabilities of data storage and visualisation of the changes, activities, periods and planned and actual sequences, is likely to make the dispute resolution process more efficient.

The potential of BIM in respect of dispute avoidance and dispute resolution is presently an almost entirely untapped resource. However, it is clear from the above examples, and in view of the anticipated further developments of the BIM process and technology generally, that BIM technology may play a major role in both avoiding and resolving disputes.

Virtual models will be able to help support a written narrative to provide better understanding of what occurred on a project and are particularly useful (on the 'a picture paints a thousand words' principle) when communicating information to judges, tribunals or others who have not been directly involved in a project and may have limited practical construction experience.

Given that virtual models are required on all BIM projects, it is likely that there will be an increase in their use to support claims and disputes in the future.

There are already examples where disputes have been settled with the help of BIM technology.

One such case where BIM was used for forensic investigation purposes was in respect of the collapse of the I-35W Mississippi River bridge in Minnesota in 2007. BIM assisted the parties to catalogue and access existing information on every element of the truss bridge and enabled communication between the clients and investigation team.

Another case where BIM was used in forensic investigations was the façade examination of the Metrodome roof deflation in Minnesota. The challenge of collecting and managing the substantial amount of historical and new data in the project was the main motive for implementing BIM in façade examinations.

Consideration should be given to incorporating into appointments a project-specific dispute resolution plan addendum to be utilised on BIM-enabled projects to help solve disputes in a quick, goal-orientated, proper and professional manner. This addendum may reduce the time limit for the decision to be made regarding a dispute referred because of the incorporation of digital technology. BIM acts as a centralised electronic hub, ensuring up-to-date information that is promptly available to all parties involved in the project; therefore, when a claim arises, all the required information can be accessed, filtered, sorted and reported faster than traditional paper-based or unmanaged electronic information.

11.2. Use of models

BIM models can play a large role in considering extension of time claims and disruption claims (by re-winding the model to the relevant time).

While disruption claims are common in construction and infrastructure disputes, they are still perceived by many to be 'difficult', with the burden associated with demonstrating the impact of disruption often being difficult to achieve.

A particular problem with disruption claims is proving the scale of the disruption said to have been suffered. A contractor may believe with absolute certainty that a project has suffered from indirect, knock-on impacts of disruption events for which the appointing party is responsible, but may struggle to evidence this knock-on impact with project records.

While there is much commentary around this area (including reference to approaches such as the 'measured-mile' approach – an approach that attempts to compare productivity pre-disruption with that post-disruption), what constitutes an acceptable method of presenting a disruption claim is still open to question.

As the second edition of the Society of Construction Law (SCL) *Delay and Disruption Protocol* (SCL, 2017) acknowledges, there is a wide variety of methods of presenting a disruption claim presently in use, including project-specific studies, project comparison studies and industry-wide studies.

One method of disruption analysis that is growing in popularity (even though it also has its fair share of critics) is system dynamics (SD) modelling, which is easily relatable to the BIM process. SD modelling is referred to in the second edition of the SCL *Delay and Disruption Protocol* (SCL, 2017).

Where most disruption claims are based on quantitative analysis of project records, SD modelling sidesteps the 'real' project record and creates instead a computer simulation of the construction project, using software that allows various 'but-for' scenarios to be simulated to postulate the impact of employer-responsible disruption. The model reproduces the actual labour hour expenditures (including the as-built programme and added variations and other changes). The project is then re-simulated in the absence of the claim items resulting in a 'but-for' model. The robustness of the conclusions derived from this analysis is obviously dependent on: (*a*) the accuracy and completeness of the source input data and hence the quality and availability of project records; (*b*) the reasonableness (and impartiality) of the analyst's judgements in establishing the model and (*c*) the transparency of the analytical process carried out by the specialist software.

Despite the above issues, the SD analysis approach may be attractive to some contractors who, for whatever reason, do not feel that their documentary evidence would sustain a disruption claim using a more traditional methodology, while, at the same time, appointing parties are likely to highlight the fact that the analysis is (in large parts) theoretical in nature and is therefore divorced from reality.

Although there is little in the way of court cases that have considered SD as evidence, there is some anecdotal evidence suggesting that SD modelling is popular with some claimants. As matters progress it seems inevitable (particularly on BIM projects) that those involved in the construction industry will encounter (and will be tempted to use) SD models as evidence in dispute proceedings.

Although SD modelling was first used 'prospectively' to model alternative hypotheticals affecting ongoing projects, SD experts now offer 'retrospective' modelling services for dispute resolution.

SD studies of construction projects have concluded that small-scale disruption events are capable of causing very significant overall disruption, and this means that, in the context of a disruption claim, SD models of construction projects may often (although not always) attribute much greater 'indirect' disruption to a claim event than the 'direct' impact of that event might suggest.

The results of an SD analysis are usually presented as graphical curves of cumulative data over time. The difference between the results of the 'as-built' simulation and the 'but-for a specific claim event' simulation constitutes the SD model's assessment of the disruption impact flowing from that event.

It must be understood that, by definition, SD does not seek to produce an exact replica of the project it models; it merely seeks to arrive at a realistic model of the project. However, SD experts will typically undertake a process of validation to ensure that the SD model does not contradict what is known about the project, thereby attempting to tie it back to known facts and dates.

The initial hurdle which an SD model must overcome is to be admitted as evidence. In England and Wales, the court has discretion whether to permit expert evidence. Civil Procedure Rule (CPR) 35.1 restricts expert evidence to 'that which is reasonably required to resolve the proceedings' (HMG, 2023). CPR 32.1(2) permits the court to exclude evidence which would otherwise be admissible (HMG, 2023). In assessing what expert evidence is reasonable, the court will have regard to whether the benefits of that evidence justify the costs. The court must therefore be satisfied that the SD analysis is sufficiently useful in determining the disruption claim and that the costs of producing it are justified.

The fact that SD modelling is relatively unknown in the English courts is likely to increase the level of scrutiny with which the Technology and Construction Court (TCC) may approach an application to adduce SD expert evidence. However, particularly as the influential SCL *Delay and Disruption Protocol* (SCL, 2017) now refers to SD modelling, it is reasonable to consider that it is unlikely that the TCC would now dismiss an application to adduce evidence simply for the reason that the evidence is an SD model.

The fairly recent TCC case of *Amey LG* v. *Cumbria County Council* (2016) may be an indication as to whether a court is likely to admit an SD analysis as evidence. In that case, the court held that it was permissible to adduce non-probability sampling as evidence in support of an employer's counterclaim for defects, although it must be noted that it was also found that it was not practicable or cost-effective for the employer to have identified every instance of defect.

There would appear to be a reasonable likelihood that a contractor seeking to rely on an SD analysis in the English courts would be able to put the SD analysis before the court prior to the decision on its admissibility being made.

There is an obvious tension between, on the one hand, the claims of SD experts that an SD analysis/model knows more about the project than a witness of fact and, on the other hand, the importance of witnesses of fact in adversarial dispute resolution being the ultimate authorities about what actually happened on the project. This is a fundamental tension because it goes to the separate 'domains' of fact and expert evidence; if this tension if not overcome, it will severely weaken the credibility of SD evidence.

The SCL *Delay and Disruption Protocol* (SCL, 2017) gives the impression that an SD model is always presented to the court with a single 'but-for' simulation which indicates the number of additional labour hours attributable to the totality of employer-responsible disruption.

Although this is sometimes done, it is suggested that a better approach is for the SD expert to run a variety of alternative simulations in the model to show the individual impact of each separate claim event. If the SD expert does not do this, the court will not be able to rely on the model in the event that the employer is found liable for some of the disruption events but not others. Modelling claim events individually will give the SD model the same degree of utility to the court as a delay and disruption analysis, which allows the court to identify responsibility for individual delay and disruption events and then rely on the assessment of delay and disruption impact in whichever delay and disruption analysis the court favours.

Where a contractor presents a traditional delay analysis in support of a claim for extension of time alongside an SD model in support of a disruption claim, care needs to be taken that any discrepancies between the 'delay results' of the SD model and the delay analysis are not greater than would be expected given the SD model's level of aggregation. The structure of an SD model of a construction project should be informed where possible by the activity relationships in the contractor's time schedule, and the 'as-built' results of the SD model should be compared against schedule updates and progress records.

Although there are many challenges in bringing a disruption claim based on an SD model, in principle there is no reason why this cannot be done successfully. A properly constructed SD model is rooted in the project record and the testimony of witnesses of fact. SD modelling is based on extensive academic research into the feedback relationships within systems and projects, including construction projects. There is a 'synergy' between the problems commonly acknowledged in proving the scale of disruption and the solutions offered by SD modelling of project feedback relationships. It seems likely that more and more contractors will bring claims using SD models.

11.3. Artificial intelligence (AI)

As we all know, artificial intelligence (AI) is progressing rapidly. Artificial intelligence focuses on making already 'intelligent' systems capable of simulating human-like decision making and execution, enabling those systems to perform functions traditionally executed by skilled human professionals but at a much higher level because of the speed and power available on modern computing platforms.

AI systems can be very simple, such as business rules that replicate a simple human decision, or extremely complex, such as executing real-time customer 'conversations' in a call centre, while still providing a natural experience to the customer.

Advances in data processing speeds, lower costs, big data volumes and the integration of data science into technology has made AI practical for many organisations and has brought it within the reach of many more organisations.

Although AI has generated a huge amount of interest generally, including in the legal world, the extent of the interest has not necessarily reflected the extent to which AI is actually being used. However, that position is quickly changing as parties become more confident with the use of AI and as appropriately mature tools become more readily available in the marketplace.

AI can be used in respect of the physical built assets themselves; for example, it is now quite common for miniature wireless ultra-low power sensors to be combined with AI for structural monitoring of infrastructure assets and so on. Modern technology offers AI that is capable of being self-calibrating and of analysing data in real time. The installed sensors enable surveyors to remotely monitor the condition and structural health of the infrastructure asset remotely without having to enter potentially unsafe sites to carry out physical inspections.

In terms of administration matters, AI chatbots with natural language processing that allow human-like conversations to take place to complete various tasks are becoming more and more common. These generative AI tools can, among many other things, answer questions and assist with composing written documents. There is plenty of anecdotal evidence that legal professionals use AI chatbots on a regular basis to do some of the mundane background work relating to legal submissions and the like.

However, it must be kept in mind that the AI technologies have flaws and that generative AI tools are prone to the possibility of creating false citations or facts and of amplifying existing biases present in the underlying data set.

A cautionary tale came in June 2023, where two New York attorneys and their firm were fined $5000 for misleading a New York court after filing a brief referring to non-existent case law generated by a generative AI tool. While legal practitioners may be sanctioned if they fall short of their duties, algorithms (which generative AI tools rely on) obviously have no such obligations. The judge in the New York case noted that ethics rules applicable to the lawyers imposed a 'gatekeeping role' on them to ensure the accuracy of their filings, including their use of AI (Merken, 2023). At the present time, AI tools cannot yet be relied on without careful human review.

Some US judges have issued standing orders trying to grapple with this problem, requiring disclosure if AI has been used in drafting pleadings as well as certification that the accuracy of AI pleadings have been verified. Other US courts have issued more onerous orders requiring disclosure of the tool and manner of use of AI in legal research and in drafting any documents for filing. Canadian courts have also issued general practice directions requiring disclosure of the use of AI and the manner of its use in any drafting or legal research. In US and Canadian courts, case management powers have so far been found to be sufficient to regulate and sanction lawyers' use of AI, but this is likely to change in the future.

While the US and Canadian court orders and practice directions have been very wide, often requiring disclosure of all uses of AI in research or court filings, it is in reality very difficult to know whether AI has been used at all. Tools exist to detect the use of AI but, as the technology becomes increasingly sophisticated, it may become almost impossible to know whether drafts are prepared by AI or by humans, and it is possible to imagine a scenario where failure to disclose the use of AI is almost undetectable.

In response to this issue, in the UK, guidance for responsible use of AI in courts and tribunals was published by the Courts and Tribunals Judiciary in December 2023 (Courts and Tribunals Judiciary, 2023). This guidance made it clear that, before using any AI tools, the person proposing to use the AI must have a basic understanding of the AI capabilities and the potential limitations.

Some of the key limitations listed are as follows.

- Public AI chatbots do not provide answers from authoritative databases. They generate new text using an algorithm based on the prompts they receive and the data they have been trained on. This means the output which AI chatbots generate is what the model predicts to be the most likely combination of words (based on the documents and data that it holds as source information). It is not necessarily the most accurate answer.
- As with any other information available on the internet in general, AI tools may be useful to find material you would recognise as correct but have not got to hand, but are a poor way of conducting research to find new information you cannot verify. They may at best be seen as a way of obtaining non-definitive confirmation of something, rather than providing immediately correct facts.
- The quality of any answers received from the AI chatbot will depend on how a user engages with the relevant AI chatbot, including the nature of the prompts that are entered. However, even with the best prompts, the information provided may be inaccurate, incomplete, misleading or biased.

Another issue raised is in respect of confidentiality and privacy. A checklist of important issues is provided below.

- Users should not enter any information into a public AI chatbot that is not already in the public domain.
- Users should not enter any information into a public AI chatbot which is private or confidential.
- Any information that is input into a public AI chatbot should be seen as being published to all the world.
- The current publicly available AI chatbots remember every question asked of them, as well as any other information you put into them. That information is then available to be used to respond to queries from other users. As a result, anything a user types into a public AI chatbot could become publicly known.
- Some AI platforms, particularly if used as an app on a smartphone, may request various permissions which give them access to information on the user's device. In those circumstances, you should refuse all such permissions.
- In future, AI tools designed for use in the courts and tribunals may become available but, until that happens, users should treat all AI tools as being capable of making public anything entered into them.

It is vital that a user ensures that the AI information obtained is accountable and accurate.

- The accuracy of any information that a user has been provided with by an AI tool must be checked before it is used or relied on.
- Information provided by AI tools may be inaccurate, incomplete, misleading or out of date, and, even if it purports to represent English law, it may not do so.
- AI tools may make up fictitious cases, citations or quotes; refer to legislation, articles or legal texts that do not exist; provide incorrect or misleading information regarding the law or how it might apply; and make simple factual errors.

Users must be aware of potential bias as AI tools generate responses based on the dataset they are trained on; therefore, information generated by AI will inevitably reflect errors and biases in its training data. A user should always have regard to this possibility and the need to correct this.

Possibly one of the largest issues is the matter of security. Therefore, to maintain security, users should follow best practices for maintaining their own and the court/tribunals' security, including the following.

- Users should use work devices (rather than personal devices) to access AI tools.
- Users should use work email addresses.
- A paid subscription to an AI platform should be used where possible as paid subscriptions have been identified as generally more secure than non-paid access. However, be aware that there are a number of third-party companies that license AI platforms from others and these are not as reliable in how they may use information that is input.

Users need to take responsibility for data obtained, keeping in mind that judicial office holders are personally responsible for material which is produced in their name but that judges are not generally obliged to describe the research or preparatory work which may have been done in order to produce a judgment.

People need to be aware that court/tribunal users may have used AI tools; some kinds of AI tools have been used by legal professionals for a significant time without difficulty. For example, TAR (technology assisted review) is now part of the landscape of approaches to electronic disclosure.

Unfortunately, AI tools are now being used to produce fake material, including text, images and video. Courts and tribunals have always had to handle forgeries, and allegations of forgery, involving varying levels of sophistication; judges and tribunals should be aware of this new possibility and potential challenges posed by deepfake technology.

Therefore, in summary of the use of artificial intelligence (AI) in legal submissions, there are certain pros and cons as follows.

Pros

- access to justice resolving disputes on a cost-effective and timely basis
- more emphasis on outcomes rather than process
- proportionality in terms of cost and time
- predictability: one of the key properly functioning aspects of law is the predictability of the outcome
- transparency: there is no query about how a decision was arrived at, but there may still be reasons to disagree with the decision.

Cons

- a lack of humanity and dignity in decision making
- contrary to the above list, a lack of transparency because of reliance on unknown and uncertain data

- the devolution of standards to rules: the essence of law with its flexibility and sensitivity to adjust for new circumstances will be forsaken for expediency
- the lack of the calcification and evolution of the law – in other words, how will a system that does not think or reason or have any sense of human experience have any legitimacy in developing the law?
- the lack of accountability and liability, whereas judges who get it wrong can be reprimanded
- absence of empirical evidence – that is, one hat does not fit all
- due process concerns – that is, not sufficient safeguards for being given a fair opportunity to present and/or defend a case
- demise of legal institutions
- de-skilling of humans
- inability to find the required facts – technology cannot necessarily do that
- unacceptability of AI interpreting and evolving the law.

There is a school of thought that AI could do away with the need for courts/tribunals entirely in the future, particularly as AI can produce a reasoned, quick and cost-effective decision by simply ascertaining the facts and applying the law (which, in general terms, is the approach taken by courts and tribunals), with the added advantages (if they are advantages) that no issues of fairness and/or equity will be allowed to encroach on the decision-making process. We may be some way from that possibility at the moment, but it is something that is bound to become a more likely possibility as time passes.

11.4. Case law review

There have been relatively few court cases in the UK that have related directly to BIM, although, anecdotally, there is evidence that BIM-related issues have featured more in adjudications and in arbitrations.

Despite this, the following cases provide an indication of the types of issues that the courts have had to address.

Birse Construction v. St David (1999)

The case of *Birse Construction* v. *St David* (1999) is significant in the context of BIM due to the legal and contractual risks it highlights. In that case, Birse Construction was engaged by St David to carry out work on a development project. However, disputes arose regarding whether a binding contract had been formed, as the JCT form of contract discussed was never signed. The court ultimately concluded that no contract had been formed, which meant that St David could not rely on the arbitration clause in the unsigned contract. This case underscores the importance of having clear, binding contract terms and, with parallels to the BIM process, also underscores the need for the parties to have a clear understanding of all parties' roles and responsibilities in BIM projects. It illustrates the risks of having non-binding collaboration and the potential for misunderstandings and disputes. An optional or non-binding 'collaborative' understanding gives rise to a real risk that the collaborative promises will simply fall by the wayside if a dispute subsequently arises.

Amey v. Cumbria County Council (2016)

The TCC case of *Amey* v. *Cumbria County Council* (2016) may be an indication as to whether a court is likely to admit an SD model as evidence. HHJ Stephen Davies held that it was permissible to adduce non-probability sampling as evidence in support of an employer's counterclaim for defects.

Trant Engineering v. Mott MacDonald (2017)

As discussed earlier in Chapter 5, the hosting of the CDE was an issue in the *Trant Engineering* v. *Mott MacDonald* (2017) TCC court case. As a result of a dispute, Mott MacDonald had blocked Trant's access codes to the CDE. The TCC ordered that Mott MacDonald should restore access to the relevant design materials. Please see Chapter 5.6 for more details.

Balfour Beatty v. Broadway Malyan (2022)

Balfour Beatty's failed legal attempt to compel Broadway Malyan to retrieve documents relating to a building completed 13 years previously highlights not only the issues that information management can address in construction but also the challenges around its retrospective application to the existing built environment.

11.5. Summary

- Disputes in the construction industry can have a major impact on organisations and projects.
- The implementation of BIM presents an opportunity to nip in the bud any disputes before they have the opportunity to develop and also provides contemporaneous information that should greatly assist in the resolution of a great many other disputes.
- BIM models can play a large role in considering extension of time claims (by re-winding the model to the relevant time) as well as disruption claims.
- System dynamics (SD) modelling is easily relatable to the BIM process; it is referred to in the second edition of the SCL *Delay and Disruption Protocol* (SCL, 2017).
- Artificial intelligence (AI) can be used in dispute resolution but has its shortcomings. For this reason, the Courts and Tribunals Judiciary published guidance for the responsible use of AI in courts and tribunals in the UK (Courts and Tribunals Judiciary, 2023).
- Before using AI in respect of dispute resolution, the pros and cons need to be listed and considered in detail.
- There have been relatively few court cases in the UK that have related directly to BIM, although there is anecdotal evidence that BIM-related issues have featured more in adjudications and arbitrations.

Peter Barnes
ISBN 978-1-83608-515-7
https://doi.org/10.1108/978-1-83608-512-620251013
Published under exclusive licence by Emerald Publishing Limited

Chapter 12
Conclusion

BIM is now widely adopted, both in the UK and internationally and, as time moves on and as technology improves and develops, BIM models will give more and more direct access to the stored construction documents through links to the appropriate server; even now, the digital flow of information from inception through to demolition is becoming more and more common.

Technological improvements are occurring frequently, but the contractual and legal consequences of using BIM are (almost inevitably) lagging behind somewhat. It is vitally important that the legal and contractual issues relating to BIM are taken into consideration when operating under BIM: navigating the legal and contractual landscape in BIM is essential for the successful and ethically sound implementation of projects within the construction industry.

A starting point is usually the BIM strategy produced by (or on behalf of) the appointing party (sometimes referred to as the client or the end-user); this BIM strategy establishes what the appointing party wants to obtain or needs to obtain in respect of information and/or data (known as the BIM deliverables) through the use of BIM. This BIM strategy is normally developed by the appointing party (often with the assistance of an independent BIM adviser) from the appointing party's statement of need which defines and describes their outline BIM requirements and deliverables.

Following on from this, the appointing party needs to appoint one or more lead appointed parties to operate the BIM procedures and to lead teams (delivery teams) within the process. These teams are required to provide information about the built asset that the appointing party wishes to procure, in line with the exchange information requirements, normally under the headings of technical, management and commercial. A lead appointed party is appointed directly by the appointing party and takes the role of a coordinator between a delivery team and the appointing party, and can assign tasks to its own appointed parties (i.e. a party tendering for or actually appointed to a project).

To encourage standardisation and collaboration, it is important that standard BIM specifications/standards are established and used; the BS EN ISO 19650 (BSI, 2019) standard is an international set of standards of good practice for managing information over the whole lifecycle of a built asset using BIM.

At an early stage of the BIM process, the appointing party needs to establish what BIM models are required. Two of the most important models are the project information model, which is a model that is developed during the design, production and construction phase (the delivery phase) of a project, and the asset information model, which is developed from the project information model and which contains information related to the use, management, maintenance and operation of the built asset.

The BIM process comprises of eight steps (as extracted from BS EN ISO 19650-2 (BSI, 2018b)) and is normally broken down into a procurement phase, a planning phase and a production phase. As part of these processes, a mobilisation plan and a project implementation plan need to be produced. Appointments are made incorporating the post-appointment BIM execution plan, the current responsibility matrix, the applicable task information plans and the master information delivery plan.

Two key roles are those of the BIM coordinator, who essentially has a design role with responsibility for model coordination and clash detection, and the BIM information manager, who is essentially a procedural gatekeeper, policing the model(s), the common data environment (CDE) (i.e. the storage area for information and models) and the associated processes.

A very important aspect of the BIM process is the need for lessons to be learnt (for use in future BIM projects) and the 'golden thread' of information which needs to be developed.

This golden thread of information is a digital record of everything associated with a built asset, including processes, materials, decisions and usage. It is a live repository linking all asset data and its purpose is to ensure that everyone involved in building safety management has access to key details that could help identify, assess and mitigate risks. The golden thread of information enables accountable persons to demonstrate that they understand the building and the steps required to keep it safe. This is an aspect of health and safety that is increasing in importance, particularly in the light of the Building Safety Act 2022 (HMG, 2022).

Currently there is no standard market-led approach in respect of the incorporation of BIM into appointments/contracts, but most standard form contracts rely on an information protocol dealing with BIM issues being added as an addendum to the contracts.

However, as collaboration is a key element of BIM, it is apparent that more effort needs to be made to develop new contracts (possibly smart contracts and/or contracts in blockchain) which put collaboration at the heart of the BIM appointments.

In light of the above need for collaboration, the concepts of design and design liability and the position of intellectual property rights and copyright need to be considered in the context of BIM projects.

A particular area of concern is how information security threats need to be dealt with, which may increase because of the increasing use of artificial intelligence generally and in the BIM process specifically. Consequently, a structured risk management process needs to be put in place and a risk assessment needs to be carried out, always keeping in mind the CIA triad (confidentiality, integrity and availability). Also, due consideration must be given to the relevant data protection laws and, in the UK, in particular, to the General Data Protection Regulation (GDPR).

Finally, and while still very much a developing area, BIM may in the future have a major impact on the way that disputes are resolved. The implementation of BIM presents an opportunity to nip in the bud any disputes before they have the opportunity to develop, as well as providing contemporaneous information that should greatly assist in the resolution of a great many other disputes.

Therefore, all in all, while exciting times lie ahead in the increased use of the BIM process, there are many contractual and legal aspects relating to the BIM process which cannot be forgotten and which cannot be left behind if BIM is to reach its full and beneficial potential. It is hoped that this book will assist in guiding parties along the path of achieving the full benefits of BIM.

Peter Barnes
ISBN 978-1-83608-515-7
https://doi.org/10.1108/978-1-83608-512-620251017
Published under exclusive licence by Emerald Publishing Limited

References

ACA (Association of Consultant Architects) (2013) *PPC2000 (Project Partnering Contract)*. ACA, Hanworth, UK. https://ppc2000.co.uk/ppc2000 (accessed 21/11/2024).

Amey LG Ltd v. *Cumbria County Council* (2016) EWHC 2946 (TCC). https://www.bailii.org/ew/cases/EWHC/TCC/2016/2946.html (accessed 21/11/2024).

Balfour Beatty Regional Construction Ltd v. *Broadway Malyan Ltd* (2022) EWHC 2022 (TCC). https://www.bailii.org/ew/cases/EWHC/TCC/2022/2022.html (accessed 21/11/2024).

Birse Construction Ltd v. *St David Ltd* (1999) EWHC TCC 419. https://www.bailii.org/ew/cases/EWHC/TCC/1999/253.html (accessed 21/11/2024).

BSI (2018a) BS EN ISO 19650-1: Organization and digitization of information about buildings and civil engineering work, including building information modelling (BIM). Information management using building information modelling. Concepts and principles. BSI, London, UK.

BSI (2018b) BS EN ISO 19650-2: Organization and digitization of information about buildings and civil engineering work, including building information modelling (BIM). Information management using building information modelling. Delivery phase of the assets. BSI, London, UK.

BSI (2019) BS EN ISO 19650 series: Organization and digitization of information about buildings and civil engineering work, including building information modelling (BIM). Information management using building information modelling. BSI, London, UK. https://www.bsigroup.com/en-GB/capabilities/buildings-and-construction/iso-19650-building-information-modelling-bim (accessed 21/11/2024).

BSI (2020a) BS EN ISO 19650-3: Organization and digitization of information about buildings and civil engineering work, including building information modelling (BIM). Information management using building information modelling. Operational phase of the assets. BSI, London, UK.

BSI (2020b) BS EN ISO 19650-5: Organization and digitization of information about buildings and civil engineering work, including building information modelling (BIM). Information management using building information modelling. Security-minded approach to information management. BSI, London, UK.

BSI (2022a) BS EN ISO 19650-4: Organization and digitization of information about buildings and civil engineering work, including building information modelling (BIM). Information management using building information modelling. Information exchange. BSI, London, UK.

BSI (2022b) BS 8536: Design, manufacture and construction for operability. Code of practice. BSI, London, UK. https://knowledge.bsigroup.com/products/design-manufacture-and-construction-for-operability-code-of-practice?version=standard (accessed 21/11/2024).

BSI (2023) BS EN ISO 19650-6: Organization and digitization of information about buildings and civil engineering work, including building information modelling (BIM). Information management using building information modelling. Health and safety information. BSI, London, UK.

CIC (Construction Industry Council) (2018) *Building Information Modelling (BIM) Protocol*, 2nd edn. CIC, London, UK. https://www.cic.org.uk/uploads/files/old/bim-protocol-2nd-edition-2.pdf (accessed 21/11/2024).

CIOB (Chartered Institute of Building) (2013) *Contract for Use with Complex Projects*. CIOB, Bracknell, UK. https://www.ciob.org/sites/default/files/CIOB-AGREEMENT-NO1-DIGITAL-AW.pdf (accessed 21/11/2024).

CLC (Construction Leadership Council) (2016) *Modernise or Die: The Famer Review of the UK Construction Labour Model*. CLC, London, UK. https://www.cast-consultancy.com/wp-content/uploads/2021/03/Farmer-Review-1-1.pdf (accessed 16/10/2024).

Courts and Tribunals Judiciary (2023) *Artificial Intelligence (AI). Guidance for Judicial Office Holders*. https://www.judiciary.uk/guidance-and-resources/artificial-intelligence-ai-judicial-guidance/#related_content (accessed 21/11/2024).

Egan J (1998) *Rethinking Construction: Report of the Construction Task Force*. HMSO, London, UK. https://constructingexcellence.org.uk/rethinking-construction-the-egan-report (accessed 16/10/2024).

Fenice Investments Inc. v. *Jerram Falkus Construction Ltd* [2009] EWHC 3272 (TCC). https://www.bailii.org/ew/cases/EWHC/TCC/2009/3272.html (accessed 16/10/2024).

Gerbert P, Castagnino S, Rothballer C, Renz A and Filitz R (2016) *Digital in Engineering and Construction: The Transformative Power of Building Information Modeling*. BCG (Boston Consulting Group), London, UK. https://web-assets.bcg.com/img-src/BCG-Digital-in-Engineering-and-Construction-Mar-2016_tcm9-87277.pdf (accessed 21/11/2024).

HMG (Her Majesty's Government) (2020) *The Construction Playbook*. The Stationery Office, London, UK. https://www.gov.uk/government/publications/the-construction-playbook (accessed 21/11/2024).

HMG (2022) Building Safety Act 2022. The Stationery Office, London, UK. https://www.gov.uk/guidance/the-building-safety-act (accessed 16/10/2024).

HMG (2023) Civil Procedure Rules. The Stationery Office, London, UK. https://www.justice.gov.uk/courts/procedure-rules/civil/rules (accessed 21/11/24).

ICO (Information Commissioner's Office) (2016) UK General Data Protection Regulation. https://ico.org.uk/about-the-ico/what-we-do/legislation-we-cover/general-data-protection-regulation/ (accessed 21/11/2024).

Infrastructure and Projects Authority (2021) *Transforming Infrastructure Performance: Roadmap to 2030*. The Stationery Office, London, UK. https://www.gov.uk/government/publications/transforming-infrastructure-performance-roadmap-to-2030 (accessed 16/10/2024).

JCT (Joint Contracts Tribunal) (2019) *BIM and JCT Contracts*. JCT, London, UK. https://corporate.jctltd.co.uk/initiatives/bim (accessed 16/10/2024).

JCT (2024) *Design and Build Contract 2024*. JCT, London, UK.

JCT (2024) *Standard Building Contract 2024*. JCT, London, UK.

Lara V (2015) *Rethinking Productivity Across the Construction Industry*. EIU (Economist Intelligence Unit), sponsored by Autodesk. https://impact.economist.com/perspectives/strategy-leadership/rethinking-productivity-across-construction-industry (accessed 21/11/2014).

Latham M (1994) *Constructing the Team*. The Stationery Office, London, UK. https://constructingexcellence.org.uk/constructing-the-team-the-latham-report (accessed 16/10/2024).

Law Commission (2021) *Smart Legal Contracts: Advice to Government*. Law Commission, London, UK. https://lawcom.gov.uk (accessed 16/10/2024).

Merken S (2023) *New York Lawyers Sanctioned for Using Fake ChatGPT Cases in Legal Brief*. https://www.reuters.com/legal/new-york-lawyers-sanctioned-using-fake-chatgpt-cases-legal-brief-2023-06-22/ (accessed 21/11/2024).

NEC (New Engineering Contract) (2017) *NEC4 Alliance Contract*. NEC, London, UK. https://www.neccontract.com/products/contracts/nec4/alliance-contract/alc (accessed 16/10/2024).

NEC (2017) *The NEC4 Engineering and Construction Contract*. NEC, London, UK. https://www.neccontract.com/products/contracts/nec4/engineering-and-construction-contract/ecc (accessed 16/10/2024).

SCL (Society of Construction Law) (2017) *SCL Delay and Disruption Protocol*, 2nd edn. SCL, London, UK. https://www.scl.org.uk/sites/default/files/documents/SCL_Delay_Protocol_2nd_Edition_Final.pdf (accessed 16/10/2024).

Trant Engineering Limited v. *Mott MacDonald Limited* [2017] EWHC 2061 (TCC). https://www.fenwickelliott.com/research-insight/annual-review/2017/uk-bim-trant-mott-macdonald (accessed 16/10/2024).

UK BIM Framework (2020) *Information Protocol to Support BS EN ISO 19650-2: The Delivery Phase of Assets*. UK BIM Framework, London, UK. Information-Protocol-to-support-BS-EN-ISO19650-2.pdf (ukbimframework.org) (accessed 16/10/2024).

UK NAO (UK National Audit Office) (2001) *Modernising Construction*. The Stationery Office, London, UK. https://www.nao.org.uk/reports/modernising-construction (accessed 16/10/2024).

FURTHER READING AND USEFUL WEBSITES

ASCE (American Society of Civil Engineers) (2013) A new method for managing and presenting façade investigation data. *ASCE Civil Engineering Magazine* **83(1)**.

ASCE (2013) Forensic information modeling: A new forensic tool. *ASCE Civil Engineering Magazine* **83(1)**.

BIM4Legal (now known as IM4Legal). https://wearenima.im

Bitcoin. https://bitcoin.org/en

Buildingsmart Data Dictionary. https://www.buildingsmart.org

Centre for Digital Britain. https://www.cdbb.cam.ac.uk

CIC (Construction Industry Council). https://www.cic.org.uk

CIOB (Chartered Institute of Building). https://www.ciob.org

Civil Procedure Rules. https://www.justice.gov.uk/courts/procedure-rules/civil

COBie (Construction Operations Building Information Exchange). https://www.thenbs.com/knowledge/what-is-cobie (accessed 16/10/2024).

Copyright, Designs and Patents Act 1988. https://www.legislation.gov.uk/ukpga/1988

Courts and Tribunals Judiciary. https://www.judiciary.uk

Data Protection Act 2018. https://www.gov.uk/data-protection

Departments for Communities and Local Government, and for Business, Energy and Industrial Strategy. https://www.gov.uk/government/organisations

GDPR (General Data Protection Regulation). https://www.gov.uk/data-protection

Global BIM Network. https://globalbim.org

Golden thread. https://www.gov.uk/guidance/keeping-information-about-a-higher-risk-building-the-golden-thread

Information protocol template, developed by the UK BIM Framework. https://ukbim-framework.org/wp-content/uploads/2020/06/Information-Protocol-to-support-BS-EN-ISO19650-2.pdf

JCT. https://www.jctltd.co.uk

NBS National BIM Library (now known as NBS BIM Library). https://source.thenbs.com/bimlibrary

PPC2000 (Project Partnering Contract). https://ppc2000.co.uk/ppc2000

Sale of Goods Act 1979. https://www.legislation.gov.uk/ukpga/1979/54

Supply of Goods and Services Act 1982. https://www.legislation.gov.uk/ukpga/1982/29

System dynamics modelling. https://systemdynamics.org

Technology and Construction Court (TCC). https://www.gov.uk/courts-tribunals/technology-and-construction-court

UK BIM Alliance (now known as NIMA). https://wearenima.im

Uniclass. https://www.thenbs.com/our-tools/uniclass

Peter Barnes
ISBN 978-1-83608-515-7
https://doi.org/10.1108/978-1-83608-512-620251015
Published under exclusive licence by Emerald Publishing Limited

Index

Printed and bound by CPI Group (UK) Ltd, Croydon, CR0 4YY
13/05/2025
14669464-0001